John Taylor is Professor of Mathematic
is well known as a writer and radio and
stimulating books which include *The S*
Physics and *Black Holes: The End of the*

Professor Taylor has taught and researched at several universities on both sides of the Atlantic, and has recently been Professor of Physics at Rutgers University, New Jersey; Faculty Lecturer at the Mathematical Institute, Oxford; Senior Lecturer in Physics and Reader in Fields and Particles at Queen Mary College, London; and Professor of Physics at Southampton University. He is a Fellow of the Cambridge Philosophical Society, the Institute of Physics and the Physical Society.

Apart from his deep scientific interest in the paranormal, he has specialized in elementary particle physics, cosmology and brain research, and is generally concerned to promote a wider understanding of the new frontiers of science.

Professor John Taylor

SUPERMINDS

an investigation into the paranormal

published by Pan Books in association with Macmillan London

First published 1975 by Macmillan London Ltd
This revised edition published in Picador 1976 by Pan Books Ltd,
Cavaye Place, London SW10 9PG,
in association with Macmillan London Ltd

Diagrams by Osborne Marks Associates
ISBN 0 330 24705 0
Printed and bound in Great Britain by
Richard Clay (The Chaucer Press) Ltd, Bungay, Suffolk

Contents

Introduction

Paranormal phenomena have long presented a serious challenge to science. Over the past hundred years various scientists have attempted to come to grips with the phenomena, but with no clear-cut success. In the last two years a quickening of interest has come with the appearance of numerous people claiming special powers, the 'superminds' of the title. This book is an attempt to describe these new developments and how established science might face up to them. Besides presenting the facts of the situation as accurately as possible it suggests a scientific explanation which allows feasibility to be attached to some of the phenomena that have occurred.

The relevance of the paranormal in understanding the nature of man and life is clear. Because of its importance the subject has a strong emotional impact, and it is surrounded by a deal of controversy. Undoubtedly the contents of this book will arouse further discussion. It has been written with the desire to present the facts and ideas noted by myself in the course of active investigation into the area over the last two years. I would have felt it only cowardly to have refused to disclose the information and the understanding I have gained. I hope it will be read in the spirit of rational inquiry in which it was written; only in that way can clarity be retained.

The investigation was time-consuming not only for myself, but also for those with whom I have worked. I would like to take this opportunity to thank those subjects whom I studied for their patience and cooperation, and to add very grateful thanks to parents for the hospitality extended to myself, colleagues and pieces of apparatus. I would also like to thank my colleagues at King's College, London, and elsewhere for their help in time and apparatus at various stages of the investigation.

King's College, London
January 1975

1 The record of events

On the evening of Friday, 23 November 1973, British television viewers were surprised and baffled by the incredible sight of a fork collapsing under the influence of Uri Geller, an entertainer from Israel. Not only did he cause the fork to break after a minute's gentle stroking, but he was able, without touching it, to cause the bending of another one lying nearby on the table. He followed this up by restarting a stopped watch, bending the second hand inside the watchglass, and capped his performance by correctly guessing the details of a drawing sealed in an envelope.

As the transmission (a Dimbleby Talk-In programme) ended, uproar broke out among the specially selected audience. People surged forward to scrutinize the bent or broken cutlery and the ticking watch and to compare Geller's sketch with the original drawing. Some sceptics protested that it was all a fraud: one, a professional magician, claimed that given time he could perform all these 'tricks'. Others, already convinced of the existence of extra-sensory powers, used the demonstration as a springboard for their own pet theories – a few even went so far as to press their pamphlets or books about the supernatural on other members of the audience.

Geller's demonstration of his amazing powers produced other effects which were even more incredible than those he had achieved himself. For it turned out that in hundreds of homes throughout Great Britain cutlery had been bent and timepieces, long defunct, restarted. Several hundred people telephoned the BBC to describe their experiences; many others wrote in. And months later, at the time of writing, the repercussions of these events had not ceased.

The extraordinarily vociferous response has persisted ever since, and has usually been polarized to the extremes of the charge of outright fraud, and of total disbelief. Geller made a number of other public appearances in Britain after this Dimbleby programme, and has on occasion since returned to this country. He has also demonstrated his powers in many other countries, including Norway, Sweden, France, Germany, Japan, and the United States. Wherever he travels he always evokes a strong response.

The strength of people's reactions to such phenomena is only natural: they do not appear to be possible to achieve by any method known to us. It seems that only trickery or unknown powers could account for them. Telepathy – the power of sending information from one mind to another without the aid of normal senses – has long been argued over, and evidence for it is slowly accumulating. But no happening has ever been so dramatic as that shown by Uri Geller to millions of British viewers on that momentous evening. No methods known to science can explain his revelation of that drawing in the envelope. And similarly startling was the metal-bending. Here was a phenomenon associated with a form of matter thought to hide far fewer secrets than the mind and brain. This bending of metal is demonstrably reproducible, happening almost wherever Geller wills. Furthermore, it can apparently be transmitted to other places – even hundreds of miles away. The Geller phenomenon – Geller's manifestations of extrasensory powers, telepathy, metal-bending and watch-starting – has remained news ever since.

The violence of the argument has not necessarily thrown any more light on the crucial questions: *what* was actually happening, and *how* it was achieved? The 'what' and the 'how' are held to be the province of scientific analysis and only through studies at a suitable scientific level can these questions be satisfactorily resolved.

The importance of the findings of such a quest are undeniable – telepathic communication across vast distances, the discovery of new sources of power through metal-bending and of the cold-working of metals and other materials. They may even, some people argue, unlock a whole new view on the Universe. Yet scientists have not proved over-anxious to try to find the 'what' and the 'how' of the Geller phenomenon. They are wary of analysing events susceptible to way-out interpretations, as extra-sensory phenomena have so often been in the past, and are especially cautious of investigating events which appear, on the face of it, to be impossible without fraud. The amount of trickery and deception, even of self-deception, in the matter of extrasensory perception, or ESP, is legendary.

Scientific analysis of ESP in general has a history stretching back at least one hundred years, to the beginnings of the Society for Psychical Research (the SPR) in London. But only a few Western scientists, albeit some of the best of their day, have actively involved themselves. Among them we find the noted chemist Sir William Crookes, the famous physicist Lord Rayleigh, and Sir William Barrett, Professor of Physics at the Royal College of Science in Dublin. Yet nowadays such research is being actively pursued in Russia and in other Iron Curtain countries and this is slowly stimulating the curiosity of scientists in the United States and Europe.

Interest in the Geller phenomenon – in particular the metal-bending – has a far shorter history. Such control over matter had not even been known to exist a decade ago, at least not in the form expressed by Geller. He himself has only been demonstrating it publicly for the last six years. The scientific community in the West has only known of its existence for about two years; it was not until the Dimbleby programme that scientists in Europe began to grasp the problems posed by this phenomenon.

Naturally enough, various explanations other than those of outright fraud have recently been suggested. But all ideas claiming a serious scientific basis have so far been rather easy to demolish. Only those involving extra-terrestrial beings or similarly unknown forces have withstood criticism – basically because factors such as these, so far removed from our understanding, are in themselves still almost equivalent to miracles.

Nor does the easy explanation that the whole situation is based on deceit stand up before the facts. Geller has caused many metallic objects to be bent or broken, sometimes without direct contact or even any chance of contact, as occurred in people's homes during the Dimbleby show. Many people have themselves observed the metal being bent. It is difficult to believe that in all cases the observers have been deceived or are involved in deception.

There are, in any case, other people who have powers similar to those of Geller, many of them children – the youngest is only seven years of age. Some of these children can bend metal without touching it; it is hard to understand how at that age they could have developed the necessary conjuring skill. Taking into account all the other evidence for ESP amassed over the last hundred years, it would seem that here there is a *prima facie* case to consider.

Uri Geller appears to have posed a serious challenge for modern scientists. Either a satisfactory explanation must be given for his phenomenon within the framework of accepted scientific knowledge, or science will be found seriously wanting. Since such an explanation appears to some to be impossible, either now or in the future, they argue that the Geller phenomenon is incompatible with scientific truth, and that the value of reason and the scientific point of view is therefore an illusion. Will the gates of unreason then be allowed to open and drown us in a world inhabited by aetheric bodies, extra-terrestrial visitors, spirits of the dead and the like? Will reason then wholly give way to superstition?

These were the anxieties aroused in me as a scientist as I sat in the studio beside Uri Geller during that telecast. Here was I, a trained physicist, a researcher for the past twenty years into the mysteries of matter and mind, witnessing something which I knew I could not explain to myself – let alone to those millions of viewers. My previous areas of research have included elementary particles, the basic constituents of matter; black holes, those 'heavy stars' which have collapsed in on themselves and appear black because they totally absorb any light shone on them; the forces of nature and, more recently, certain aspects of brain function. And here, with Geller, we could be seeing a demonstration that would, as I saw it, confound our fundamental ideas about both the latter.

Part of the job of looking at the world through the eyes of a trained scientist is to look for ways of improving our current scientific models of the world by testing them under the severest conditions. Our theories of gravity are tested to their utmost by events at the centre of a black hole where the force of gravity becomes infinite; ideas about the microstructure of matter receive the ultimate test by the scattering of elementary particles away from each other at the highest available energies. The questions raised by the Geller phenomenon appear to be of a similar, perhaps of an even more difficult nature. Because of this, there could be much to gain from a scientific explanation of the powers of Uri Geller. Above all, scientists should not shirk this challenge to the way they view the world. Ways must therefore be found of applying the scientific method to discovering the cause of the phenomenon. It is to this end that I write this book, and in particular to answer the questions: *What* is the Geller phenomenon and *how* is it to be scientifically interpreted?

The initial stage of a scientific investigation is necessarily descriptive. It is essential to go through this phase in order to determine the main features of the events under consideration, and before further study can take place with specially designed apparatus. Because people react in different ways to these incredible phenomena – sometimes becoming very disturbed – their descriptions may well become distorted and need careful corroboration.

The first step in assembling a full description of the Geller phenomenon is naturally to state what is known of the powers of Uri Geller himself. Geller was born in 1946, and he says that he first noticed his powers at the age of three, when he found he could tell his mother exactly how much she had won or lost when she came home from playing cards. At the age of seven he found that he could cause the hands of his wristwatch to jump ahead to a different hour, and later discovered that he could even cause the hands of the watch to bend. It was not till after the Arab-Israeli war of 1967 that he discovered his telepathic powers at a party where all the guests were showing off their talents. 'I decided to show what I could do. I had someone go into the next room and draw a picture, and then I copied the drawing exactly. Then I bent a key without touching it.'

Three weeks later Geller was, in his own words, 'known all over Israel, and agents were asking to book me into big halls.' Since that time he has been principally an entertainer, first in Israel, then in the United States and latterly in Europe. This was to be expected of a young man bent on enjoying the good life, but it has meant that his powers are more suspect than if he had no financial stake in his success. However, someone who has performed in public more than fourteen hundred times without once being caught cheating has some claim to be taken seriously.

Of the powers demonstrated during his act, one is an ability to determine what another person is thinking or has represented on paper. He may attempt to guess a picture that somebody has drawn and kept hidden from him. Take, for example, what happened on the evening of 23 November 1973. Several hours before the Dimbleby Talk-In, a BBC employee working on the show made an outline drawing of a yacht. This was placed in an envelope which was in turn enclosed inside another envelope, so that it was not possible to obtain any indication from the outside of what the picture was. On the programme, Geller asked the girl who had drawn the picture to concentrate on it, and after several minutes drew what he thought she had been thinking

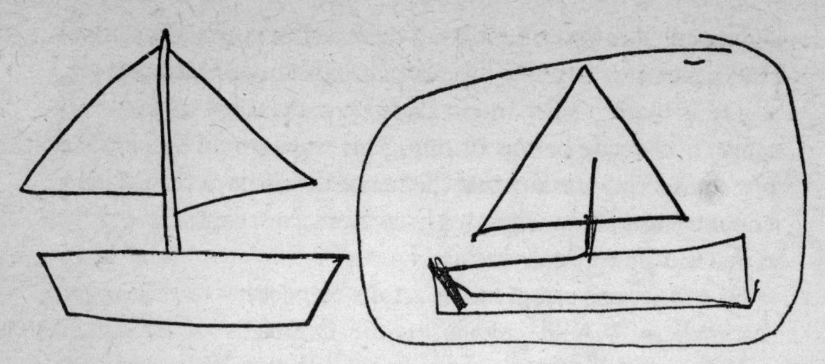

On the Dimbleby Talk-In of 23 November 1973, Uri Geller showed his telepathic powers by almost exactly reproducing a drawing of a yacht. The original had previously been drawn by a BBC employee and placed inside two envelopes (Professor John Taylor, courtesy BBC)

about. His rendering was almost an exact replica of the original, but it was framed by a line which, Uri Geller says, is the edge of the 'television screen' and was how the picture appeared in his mind. It was a remarkable feat. On 26 November 1973, Geller had similar success in the British children's television programme Blue Peter. A picture of a smiling face had been secretly drawn by the programme presenter, and Geller reproduced it almost exactly.

Another approach of his is to work with colours and capital cities. On 7 February 1974, he appeared at the New London Theatre. While Geller looked the other way, a member of the audience wrote 'Oslo' on a blackboard; an additional blackboard was placed between them. He guessed 'Oslo' accurately. He identified the colour green, although he failed with the second colour – gold. In these 'mind-reading' acts, guessing names of colours or capital cities, he is said to have an average success rate of about seventy-five per cent.

A particularly interesting case occurred at a college in Millerstown, a small Pennsylvania town. A girl student, asked to write on a blackboard the name of a capital city, wrote 'Harrisburg', the state capital of Pennsylvania, while Geller turned his back. After concentrating for several minutes Geller looked puzzled and asked, 'Are you sure this is a capital? I'm getting something, but I have never heard of this place. It looks like Harrisburg – but where is that?' Geller might have been expected to know the capital of the state he was visiting. This may surprise the sceptical, but considering the fact that the use of the name as the capital of a country was also unexpected his performance could be regarded as that much more convincing.

Of course, one can get these results by trickery. But this needs accomplices. Would anyone on the Dimbleby or Blue Peter programmes have been prepared to collaborate? The two BBC assistants involved were not known to Geller before the actual transmissions and say they drew their pictures in conditions of complete secrecy and kept them in their pockets. I saw and still see no reason to disbelieve them. In any case, Geller has so often been successful with the picture-guessing experiment that deception seems even less likely: too many people have drawn pictures for Geller's accurate guess for it to have been probable that *all* were in collusion with him. I have myself drawn pictures for Geller. And I have certainly not been in collusion with Geller – indeed, I have bent over backwards to keep an open mind about ESP. It is just possible that Geller could have used the sounds or strokes of my pen as clues to the pictures: most unlikely, however, as I repeated parts of the drawings, and also made motions of my pen over other parts of the paper. Nor did it seem possible, in a strange hotel room, for him to have used specially concealed mirrors.

Among those who have most closely observed Geller's powers have been newspaper reporters, a race not generally regarded as gullible, and who know only too well how others can be manipulated. The account of a telepathic experience with Geller by Bryan Silcock, Science Correspondent of the *Sunday Times*, is particularly relevant here:

'He [a *Sunday Times* photographer] drew a circle with another wavy circle round it, while Geller turned away, hiding his eyes, and Geller then described it exactly. First, though, he said he would try to pass the shape on to me, so I made my mind a blank. While Geller was concentrating two shapes floated into my mind, an equilateral triangle and a square with a semi-circle on one of its sides. I said nothing about them, but Geller, after describing Bryan Wharton's [the photographer's] circles, said, "Oh, I also got these," and proceeded to draw an equilateral triangle and a square with a triangle on one side of it. Since I never committed my shapes to paper he could not have watched my arm or used any magician's tricks.'

No fraud here, nor any simple explanation such as picking up clues. For the pictures only existed in Silcock's mind!

The powers of telepathy and clairvoyance (the ability to perceive distant or otherwise hidden objects directly) are not recent discoveries, though only since the latter part of the last century have they been intensively investigated. The distinctly new element which Geller has contributed to the ESP debate has to do with psychokinesis, the power of mind over matter. Not that this is new, but his particular form of it is. This was dramatically brought to the public's attention during the Dimbleby Talk-In, when his gentle stroking of one end of a fork resulted in its breaking within a minute or so. Dimbleby, who was loosely holding the other end, was heard to exclaim: 'It's bending! It's cracking! I can feel it cracking! It's breaking!'

David Dimbleby was sitting between Geller and myself. I could see the fork all the time Geller was touching it; he did not *appear* to be exerting any pressure or force, certainly not enough to break it. A startling feature of the demonstration was that the end of the fork just fell off as if the neck had become plastic a moment before. The fork which was broken had been taken from a selection on a tray put in front of the participants. A few minutes later it was noticed that another of these forks had bent at right angles at its neck.

It is as metal-bender that Uri Geller subsequently became known in England. The variety of metal objects he has bent – keys, knives, forks, spoons, nails, iron rods, scissors and so on – is astounding. He has done this often and in greatly varying conditions. For example, in my office the evening before the celebrated Dimbleby show, he bent the Yale key of a colleague of mine by stroking it gently. The total angle was no more than about 30°, a small amount of bending occurring first when the key was held by my colleague, and the further deflection occurring when it was held and stroked by Geller.

These key bendings have been legion. A *News of the World* reporter, Roy Stockdill, wrote: 'I have seen a steel plate impossible to bend by hand, suddenly curl up on a table. I have seen a key that fractured without anyone going near it.' The steel plate in question was a mirror; Geller first tried to bend it by placing the reporter's hand on it and his own on top, and then rubbing the mirror. Nothing happened, whereupon they left the mirror on a table. 'Then, as we returned to the bedroom,' the reporter wrote, 'Uri suddenly grabbed my arm and pointed to the low table on the other side of the room where we had left the mirror. The mirror was moving, bending visibly from the exact centre and rocking gently on the table. It went on moving for several more minutes, bending more and more until it formed an exact V-shape.'

The case of the fractured key concerns a hotel door key. Stockdill wrote, 'Brennan [his colleague] had just collected the key . . . He left it on top of his passport on a desk in his room. Both I and Brennan are certain that Uri went nowhere near it. Yet only minutes after Brennan had laid it down he noticed that the key had broken in two.'

So many cases of bent keys have occurred at demonstrations given by Geller that it is hard to feel any doubt that they actually did happen, either by manual contact from Geller or indirectly. When Geller demonstrated in Millerstown, Penn., a number of keys were bent in people's pockets throughout the auditorium. A security guard who had been sitting in another part of the building was heard to refer ruefully to his own twisted key, saying that he could not now lock up.

Jack Lewis's report in the *Sunday Mirror* of his experience with Geller in a London hotel room, where Geller bent a key he had touched, concluded: 'Throughout the experiment I had held my own RAC key in my hand. I glanced at it and was astonished when I saw that it had become as bent as the other one' – said to have bent 'grotesquely' – '. . . So had a metal comb which had been lying on the bed.'

No simple explanation of these occurrences of metal-bending is possible. Distracting the attention of the audience so that the bending can be achieved by force while they are all 'looking the other way' hardly seems feasible, least of all in the many cases where keys and other objects were bent or broken without even being touched by Geller, or indeed by anyone else; nor does the substitution of already bent objects for the original ones, especially with highly personal things like keys. Hypnotic suggestion is an explanation which has been put forward, but hardly seems to hold up as the objects have remained bent the following day.

At this stage it is natural to ask why such powers have not been noticed before. After all, keys and knives, forks and spoons have been around for centuries. Yet there are no documented cases of these objects being bent or broken in this way. Not until Geller has anyone had occasion to report as did Roy Stockdill of the *News of the World* that, during dinner with Uri Geller on a flight from Palm Beach, Florida to New York, he had seen his companion's spoon curl up and split in two.

One is tempted to ask why, if such powers over metal exist, they have not been appreciated earlier. Some of the reasons are fairly obvious. One could say that until the last century or so, the widespread possession of metal cutlery and watches was the exception rather than the rule, so opportunities for discovery were rare. Moreover, such powers, if made public, would in earlier times undoubtedly have earned their owner a terrible death as a witch. Thousands of men, women and children were indeed hung, burnt or drowned as witches during the sixteenth and seventeenth centuries. Witchcraft is still practised by some people today and is still a focus of fear for the community; the last trial for witchcraft in England took place as recently as 1712. The public reaction to the film *The Exorcist* in 1974 showed how strongly people still feel about such matters.

A more likely explanation is that the absence of mass communication would have limited the reputation of the persons concerned. Only their immediate families could have known about them. But television and radio have allowed millions to hear about and see the actual breaking or bending of forks and watch-hands. The feedback of all this publicity has been that reports came in of far more wide-reaching effects than had originally been imagined. Not only, it seems, do metal objects get bent near Geller without being touched by him, but they evidently do so at considerable distances, to judge by the many who telephoned or wrote after the Dimbleby Talk-In. And the same thing happened on 22 November 1973 after the sound broadcast of the Jimmy Young show. Numerous listeners had found cutlery twisted out of shape, or watches going which had not been working for months or even years.

A lady from Ilford, Essex, describing what happened to her cutlery during the Dimbleby programme, said, 'I put four forks and four spoons on a table in front of the television while Uri stroked the fork. I did not really believe anything would happen. But then I looked down and saw that one fork and three spoons had twisted out of shape. An hour after the show I had another look and found that all were distorted. It was quite frightening. I tried bending one myself but I found I could not do it.' A Harrow viewer reported that a soup ladle she had been holding had buckled in her hands during the programme. Thirty miles away, in another London suburb, a gold bracelet had become distorted.

Many people have written to me personally about such occurrences. A Liverpool woman wrote that after the Dimbleby programme, 'My son (aged 17) was keen to look in the cutlery drawer in the kitchen to see if anything had been affected, and immediately brought back the whole drawer for me to inspect. At first I could see nothing out of the ordinary until my son, white-faced, pointed out a fork which had twisted at right angles and, as you can imagine, I was pretty shaken.' She emphasized there had been no opportunity for anyone to twist the fork out of shape, and that in any case neither she nor her son would dream of deliberately ruining her cutlery.

Frequent reports come from women that the rings on their fingers feel as if they are about to buckle during Geller's demonstrations, and they feel compelled to remove them as fast as possible. A woman from Essex had this sensation so strongly that she felt her rings were about to crush her fingers. The interesting thing about this case is that her sensation coincided with the very period that Geller was on the Dimbleby show, although she was not even viewing and was unaware of the programme.

Other objects bent during the programme, including a steel knitting needle (size 10) which bent to an extent of about 15° during the programme, and a further 10° during the night. This continuation of bending after an initial distortion is not uncommon. In one household living in the South of England, the wife was holding a small spoon as Geller succeeded in breaking a fork. Nothing happened to the spoon, so it was put back in its usual place in the kitchen drawer. But, 'Next morning,' wrote her husband, 'I came down to the kitchen first and opened the drawer to take out a spoon to help myself to coffee. As I was taking the spoon out of the drawer something caught my eye and there in the fork section of the drawer was one fork bent almost at right angles.' It was a stainless steel fork, too strong for their young daughter to bend, and both husband and wife swear they had not touched it since supper the night before.

On 25 November 1973, an investigation was started by the Sunday newspaper *The People* to discover how widespread this 'bending force' was. The news editor of the paper asked all their readers to surround themselves with metal things and concentrate hard on them at twelve noon precisely, when Uri Geller would be doing likewise – but in a room in Paris. Over forty people telephoned in to say that spoons and forks had bent before their very eyes. Whether the 'bending force' had actually been transmitted from Paris is not clear, but on its own terms the experiment seems to have been successful.

Another phenomenon – watch-starting – also seems to have happened in places far removed from Geller. I have had many reports from people whose watches have started, notably one from as far as Australia where the Dimbleby programme was later shown. In this particular case, four watches started out of a total of five that had previously been defunct: 'We held the watches in our hands and four of the watches started to tick – none of which had been wound – the fifth watch rattled when it was shaken, so we really didn't expect that one to work.' Another account concerned a watch inactive for the past fourteen years; it had defeated several watch-makers' attempts to mend it. 'I placed this watch on the table beside me just before Dimbleby's programme started. As you were present, you will remember that there were perhaps fifteen minutes' chat etc before the broken watch was produced – how he tried to start it on the hand of the girl from the audience and how, after two or three unsuccessful tries, he started it on the hand of the American man. Precisely at the very moment that he started that watch on the TV programme, my watch beside me also started to work!!'

I myself have seen several watches started by Geller. He never touches them, but asks someone else to take them while he holds his hands over theirs. This method is not a hundred per cent effective, nor does the watch always keep going. The watch mentioned in the last account only went for about twelve hours and then stopped once more. It is still a remarkable phenomenon, especially when accompanied by the bending of the watch hands inside the watch glass – as occurred on the Dimbleby programme and in Geller's show at the New London Theatre.

In spite of all this evidence, the charge of fraud has still been levelled by various scientists, psychologists, magicians and journalists, and numerous articles have recently appeared in newspapers and magazines purporting to give the inside story as to how the 'deceptions' have been achieved by Geller. These reports have not taken satisfactory account of careful tests carried out by scientists, both on Uri Geller and on numbers of other people who have since been discovered with similar powers. I have myself performed such investigations with both Geller and numbers of children in England over the last year, and I have been convinced that such supernormal abilities do actually occur; the details are recounted later on in this book. They support the many accounts of key- or cutlery-bending, of watch-starting or stopping, and of telepathy which I have cited. In all, the total available evidence puts the Geller phenomenon truly into the class of the supernatural. Powers beyond the normal are possessed by a select few; these are the 'superminds' of the book's title.

Other things have gone on around Geller, so reports go, that are even stranger. One of these is the movement of objects across great distances. The first time Dr Andrija Puharich, the MD and inventor who has spent a great deal of time investigating Geller, went to Israel, he discovered that he had left his movie-camera case back in New York. He mentioned this to Geller, and the following morning was awoken by an excited telephone call. It was Geller, saying that he had found a camera case in his room. 'So I rushed over,' Dr Puharich said, 'and it was the damned thing I had left locked in an equipment closet six thousand miles away in New York. It even had my markings. Furthermore when I got back to the States I unlocked the closet and the case was gone. Since then Uri has demonstrated to me a number of times that he's capable of transporting physical objects over long distances by unknown means.'

There are others who claim to have evidence of Geller's powers of transporting physical objects. There is Dr E. C. Bastin of Cambridge, who found that his set of six screwdrivers seemed to have been moved from an upstairs room in a house in Philadelphia to somewhere on the stairs, but unfortunately all had broken in the process. A decorated egg apparently flew across the Atlantic from a drawer in a London flat to Geller's New York apartment. And now we have Edgar Mitchell, the astronaut who conducted telepathy experiments on his trip to the moon, collaborating with Geller in trying to recover a camera case he left there way back in 1969.

That is by no means all. Geller claims to have left his own body in New York and travelled to Rio de Janeiro; he returned to his physical form to find a thousand-cruzeiro note in his hand! Puharich has observed Geller enter an unidentified flying object they found in the desert – 'a disc-shaped metal object with a blue light flashing on top'. Unhappily the film cartridge containing the essential record of this event was lost, having dematerialized within a few minutes.

Such events, if they are to be taken seriously, are impossible to fit into the world as we now see it. They arouse the sceptic in us. Are the select few who claim to have witnessed these things suffering from hallucinations, we ask, or are they trying to delude others? But at a more mundane level, to guess what another person has drawn or to bend metal objects or to start broken timepieces – these are in themselves very much out of the ordinary. To come to terms with them, we must keep our minds open and not allow unqualified scepticism to blinker us.

Yet we must keep our critical faculties at the ready. Having appreciated that there may be more in the world than our present theories allow for, we should not imagine that just anything can happen, so that our scientific understanding, so painfully acquired, can be discounted. Although we may have to face apparent contradictions when bringing science to bear on extrasensory phenomena, the way ahead can only lie through rigorous scientific analysis.

The Geller phenomenon – the bending of metal or sending simple visual images from one person to another – involves only a very narrow range of extrasensory experiences. The total range of inexplicable phenomena is much wider. They include poltergeists, ghosts, tape recordings of voices from beyond the grave, precognition, mediumship, clairvoyance, astrology, dowsing, out-of-the-body experiences and the materialization of objects. Indeed, one might be forgiven for wondering whether the Geller phenomenon could ever help us understand the broader spectrum of paranormal events. But that it might help is a possibility and we must not exclude it.

Polls have shown that the paranormal lurks somewhere in the view the majority of us take of the world, and it seems that a good proportion of newspaper readers make first of all for the astrology column. Many people believe in life after death; some even claim to be in direct communication with the spirits of the dead. Cabinet ministers and royalty, as well as lesser mortals, have been regular clients of mediums and astrologers. Furthermore, the gift of extrasensory perception is commonly interpreted as a sign of transcendental powers.

Many find the purely material world of science unappealing. The attraction of religion arises from man's need for the existence of something beyond surface reality. This hidden reality is required to be a reflection of ourselves and to bear the image of our own inner mental worlds. Can paranormal phenomena give us some insight into the nature of such an underlying reality?

So many hypotheses have been made about the implications of ESP for so-called reality, and words have flowed like water. In the absence of any scientific analysis we cannot shed any light on the usually murky ideas which have been put forward. We can only hope that a careful study of the Geller phenomenon will allow us to reach a better understanding of the wider range of ESP manifestations and any underlying reality that they may expose. In this book we may find that is possible, and arrive at a unified explanation of a large body of ESP phenomena. The question is, can we expect to get from this view of reality any joy for the future of mankind? I think we can.

The way towards this goal will be taken by bringing the entire resources of science to bear on the problem of the nature of the Geller phenomenon. In the process we may well discover surprising things about the interaction of mind and matter.

2 Investigation of ESP

Modern interest in spiritualism and ESP effectively began in America in 1848. It was sparked off by inexplicable rappings in the house of a farmer called Fox, who lived in the village of Hydesville, New York. The noises, which came from the family bedroom and began to disturb everyone's sleep, had caused a crowd of people to gather. On a particularly noisy evening, alarm was caused as the rappings seemed to be giving intelligent responses to questions. The messages received were alleged to be from the spirit of a pedlar who had been murdered in the house by its previous occupant. No one had known till then of such a crime, though subsequently human teeth, bone fragments and hair were found buried under the cellar floor. More human remains were discovered later.

It then emerged that the rappings centred on Fox's younger daughter Kate, and that they had travelled with her when she moved from a previous home. Eventually a public demonstration was arranged and some tests carried out on Kate and her older sister, who also appeared to be a source of the phenomena. In a subsequent investigation the girls were searched, and then stood without shoes on feather pillows. The noises continued unabated and unexplained.

Reports of the Fox sisters' powers quickly spread and they held exhibitions of them in many places. This gave spiritualism a greater following. Within five years it was estimated that there were no less than thirty thousand recognized mediums in the United States. The ability to cause rappings spread like a kind of contagion; those who came to witness the phenomena afterwards suffered comparable disturbances in their own homes. Merely to have heard of the events sometimes proved enough for the noises to break out in people's homes.

Table tilting and levitation, materializations, and vocal messages from spirits of the dead were soon added to the rappings. Then came further means of communication – the planchette and the ouija board. These had developed in England and on the Continent, though there they were not received so favourably, especially after the great British scientist Faraday had exposed table tilting as being caused by unconscious muscular movements of the people sitting round the table. This was clear after he had used a movable top on an ordinary table; the sitters' hands were laid on the movable top and this always moved before any tilting of the table occurred.

A revival of spiritualism in England began in 1859, associated with the discovery of the powers of the celebrated medium D. D. Home, who apparently could levitate his body as well as perform other remarkable feats. As in America there was a rapid spread of people claiming to have such powers, and soon audiences of thousands attended demonstrations. This popular degree of interest was not shared by scientists; the eminent biologist T. H. Huxley refused to join an investigation of ESP and his reaction was typical: 'Supposing the phenomena to be genuine – they do not interest me. If anybody endow me with the faculty of listening to the chatter of old women and curates in the nearest cathedral town, I should decline the privilege, having better things to do . . . The only good that I can see in a demonstration of the truth of "spiritualism" is to furnish an additional argument against suicide. Better live a crossing sweeper than die and be made to talk twaddle by a "medium" hired at a guinea a séance.'

But scientists soon did become involved, particularly stimulated by the accounts in the *Quarterly Journal of Science* in 1871 by Sir William Crookes of experiments he had conducted with D. D. Home and another medium. Crookes, elected a Fellow of the Royal Society at the early age of 31 and its President in 1913, reported on an experiment in which Home was held hand and foot and closely watched. A horizontal mahogany board, pivoted at one end and the other suspended from a spring balance, was caused to be depressed by an unknown agency. Both Home and the other medium were also able to cause movements of a piece of parchment stretched lightly across a circular hoop of wood, apparently without directly touching the parchment. After searching criticism by a number of his colleagues, Crookes published further results on raps, object movement, levitation of furniture and human beings, and the appearance of lights, hands and faces.

The results obtained with Home when Crookes attached a spring balance to the edge of a table and so measured the force required to tilt it were particularly significant. A force of 20 pounds was found to be needed to tilt the table after the request had been made for the table to 'be heavy'; the normal force needed was 8 pounds. Similarly, it required a pull of 43 pounds to lift the table when 'heavy' as compared to the force of 32 pounds needed to lift it from the ground under normal conditions.

Over about twenty years in the latter part of the last century, investigations were carried out with great patience by Henry Sidgwick, one of the leading Cambridge intellectuals of his day; others were performed by investigators abroad, often with the same mediums. All the results were more or less similar: phenomena occurred which were difficult to explain naturally, but which could not easily be put down to fraud. Some of the researchers were distinguished scientists who became convinced of the genuineness of these events after very careful tests.

This was the case with Professor Richet, a French physiologist, and Sir Oliver Lodge, a British physicist. They jointly investigated a Neapolitan lady called Eusapia Palladino. She could achieve table rappings and levitations, movement of various objects and appearances of lights and hands; all the while the investigators held her hands and feet. At a later test she was caught cheating; she had wriggled out of the researchers' grip without their immediate knowledge. Despite this it is not certain that the evidence obtained with her at earlier meetings must all be rejected as being based on fraud, though it should be treated with care.

For a long time fraud was the main problem for researchers in ESP, especially at séances, customarily held in near-darkness. Various mediums were caught cheating, and investigators had to try to keep up with the latest methods of deception by becoming almost as competent as professional magicians.

The total amount of early work on these phenomena was enormous. For example, 11,000 pages were published in the *Proceedings and Journal of the Society for Psychical Research* in London between 1882 and 1900. Yet none of it was either wholly convincing to modern investigators, nor did it lead to any fruitful scientific ideas. There are various reasons for this. One clearly is that the evidence was not always free of deception. Another important one was that too many researchers were passionately eager to find evidence for survival after death; there was less interest in the physical aspects of the phenomena than in the spiritual ones. So there was little attempt to relate the events to what was then known within the physical sciences, and those eminent scientists concerned with ESP did not appear to bring very much of their scientific knowledge to bear when trying to understand the strange events they had witnessed. Explanations of paranormal phenomena were produced which had little relationship to the contemporary interpretation of the physical world – the theories could not be tested by any clear-cut predictions which might falsify them. Indeed these theories seemed impossible to quantify and only helped to produce more vague and woolly ideas. That this attitude persisted is shown by Sir Oliver Lodge's statement in the Halley Stewart lectures of 1926: 'With physical eyes alone we cannot penetrate the depths of reality.'

No wonder that many hard-headed scientists found the whole area of ESP so lacking in definition as to be repellent. For example, in 1876, when Sir William Barrett spoke about some of his experimental work on some abnormal conditions of the mind before the British Association for the Advancement of Science, his talk met with ridicule and the Association refused to publish it. It is, of course, easy to be harsh on scientists who investigated ESP: some of them did realize how difficult it was to explain the phenomena in any satisfactory way. Thus Richet wrote, at the end of a list of explanations of the events which he then dismissed, 'In fine, I *believe* that future hypothesis that I cannot formulate because I do not know it.'

We should remember that science had then not developed enough to allow the implications of various alternative hypotheses about these extrasensory phenomena to be fully explored. The forces of nature were not understood, nor how they produced those properties which matter possesses. With time that understanding grew, and techniques became more sophisticated, so we must turn to more modern work and apparatus for enlightenment.

After a slight decline, interest in ESP increased in the 1920s, the terrors of the 1914–1918 war having, no doubt, produced a need for explanations of life and death other than purely physical ones. At this stage psychologists swelled the ranks of those who were carrying out careful experiments. Many telepathy tests were performed which involved determining how successful people were at guessing cards which other people were thinking about. The results were never clear-cut, however, and required statistical analysis of the sequence of guesses to see how often the number of successes could have been obtained by chance alone. One of the earliest tests of this sort was carried out by Dr J. E. Coover, a psychologist from Stanford University. He used a pack of forty cards with the court cards removed. About one hundred different senders and guessers took part, and the sender and the guesser were always apart in different rooms. They achieved 294 successes out of 10,000, while it was calculated that the number of correct guesses which could be expected if chance alone were operating was 250. The probability against this number of successes being due to chance is about 160 to 1. Although at first glance these seem reasonably high odds, the results are not conclusive evidence for the existence of telepathy.

Further successful experiments on card-guessing were later carried out by a number of other experimenters, notably Dr J. B. Rhine at Duke University, USA. In his extensive series of tests he had some phenomenal results – one subject guessed the right cards of a special pack of five different cards over twenty-five consecutive trials. A very carefully monitored set of tests was performed at Duke University in 1938 and 1939 by Doctors Pratt and Woodruff. Two experimenters were always present as witnesses; independent duplicate records were made on special foolproof paper, one set of which was locked away. Subjects selected at random had to guess twenty-five cards, and for the 2,400 runs the success rate was so high that only once in a million such experiments could the result have been obtained purely by chance.

The evidence gathered during those early years of card-guessing was formidable. Yet that evidence has been severely criticized by some experts, particularly on the grounds that the experimental conditions could not exclude involuntary whispering by the sender, nor could it be shown conclusively that fraud had not taken place. The former charge, however, is obviously inapplicable where sender and recipient were far apart.

Several investigations were made to see if the facility of card-guessing decreased with distance. In the Pearce-Pratt experiments the average number of successes when the sender and recipient were only a yard apart was 8 out of 25, 9 out of 25 when the distance was a hundred yards, and 7 out of 25 when they were 250 yards apart. The difference in the average number of successes in each case is not significant. Other tests have shown similar lack of variation, even over very long distances. Whately Carrington, a British ESP researcher, used drawings to be guessed by subjects located in various parts of Europe and the United States, and found that when the distance between sender and recipient was greater, the success rate was higher. However, none of these experiments was really conclusive; Dr Soal, a lifelong investigator of ESP, wrote in 1951, '... in the present state of our knowledge there is little justification for the oft-repeated assertion that "telepathy is independent of space or distance".'

Fraud was hardly involved in these tests, unless it was practised on a grand scale. And unconscious deceit seems even less likely, since the experiments were much less concerned with the question of survival after death than those at the end of the nineteenth century. The only other feature which deserves scrutiny is the statistical analysis. At a trivial level, criticisms were made which suggested the existence of errors in scoring, or unfair selection of more successful trials while less successful ones were ignored, whether unconsciously or deliberately. These defects can be checked by studying the complete results of the various tests, and neither factor seems to be of any significance. A check of half a million guesses showed that only ninety mistakes had been made, and a recording of 175,000 oral guesses only contained 175 mistakes.

Strong criticism has been made of the fact that the levels of significance – specifying how much importance should be attached to a certain set of successes – were chosen far too favourably. Success expected to arise only once in several hundred similar tests is not regarded by professional statisticians as proving that chance was inoperative. Nor can probability theory be used to estimate the significance of such results without great care. For example, there can be a bunching of noughts or ones together in a random sequence of those two digits, generated, for instance, by tossing a coin. At times there would be too many noughts together, at other times too many ones. Unless very long sets of digits were chosen from the sequence they would sometimes contain what might appear to be a successive number of noughts and ones. In other words, they would not seem to have been picked from a random sequence of heads and tails, even though they actually were. Such an effect requires one to look with care at the improbability of chance success in card guessing – it may be easier to achieve, at least for limited runs, than one would think.

One set of tests which gave very disturbing results was that made by Dr Soal with a medium called Basil Shackleton. Dr Soal took great care to prevent fraud or unconscious bias. The sender and the recipient, Shackleton, were in separate rooms; only the sender could see the cards. These were five in number, lined up next to each other, initially face down, and each with a picture of a giraffe, elephant, pelican, lion or zebra. Dr Soal chose a number from a random sequence of the digits 1 to 5, the chosen number indicating to the sender which card he should turn up and concentrate on. The agent shuffled the cards after fifty guesses to introduce a further degree of randomness.

In a total of nineteen sittings between the 24 January and the 21 December 1941, Shackleton scored 1,101 correct guesses out of 3,789. Remarkably enough, these successes were on the card next to be turned up, rather than on the card the sender was actually looking at. It was calculated that this success rate could only have been obtained by chance once in every 10^{35} similar experiments – 10^{35} being the very large number of one followed by thirty-five noughts! With such odds it is extremely unlikely that Shackleton was only guessing. This result is disturbing because not only does it appear to involve the faculty of clairvoyance (viewing objects from a distance), but also the power of precognition since not even the sender knew what the card was ahead of the one he was looking at.

To some, the possibility of precognition is even more unacceptable than that of clairvoyance or telepathy. On the one hand, it would appear to deny free will; while on the other, it seems to disagree completely with one of the basic tenets upon which science is based, that of causality. This is the basic feature we have all experienced, that cause always precedes effect. The window does not break before the ball is thrown at it; the house does not explode before being hit by the bomb. Every effect seems to have a cause which precedes it; that is the principle of causality. It has been tested extensively, with great accuracy, and found valid even for what happens inside the atomic nucleus. Looking into the future involves receiving information from events yet to happen. In the last analysis this requires the ability to communicate with the past, for what is future ultimately becomes present, what is present is soon the past. Paradoxes can then occur, such as the idea of someone communicating with his past so as to prevent his mother and father from ever meeting and so conceiving him. Would he then cease to exist?

Such notions make a mockery of our ordered world, and call for a second look at the evidence for precognition. Re-analysis has shown that Soal's random numbers were not, in fact, as random as they should have been: there were too many fours and fives present. Allied with this was the earlier claim by one of the senders, Mrs Albert, that she had seen Dr Soal altering some ones into fours and fives in some of the tests. Further, it has been found that the method Soal used in constructing his sequence of random numbers was different from that appearing in his published work. So we can only have serious doubts of the Shackleton-Soal evidence for precognition.

Work in this intervening period also gave good evidence for clairvoyance. One of the most impressive experiments was that of Martin and Stribic at the University of Colorado. In a set of tests extending over three years, a total of 300,000 guesses was made by various subjects. Each subject tried to guess the order of the cards in the whole pack of twenty-five after they had been shuffled. Out of his 42,000 attempts the best subject scored an average of 6.85 successes per 25 guesses. The average success rate for all the guesses was 5.83 per 25, with an almost zero probability of these results being obtained by chance. Furthermore, a check on the correlation between each guess and the preceding card showed just chance agreement – one success in five. So one can rule out any systematic error in the data.

The control of mind over matter – levitation or other movement of objects, usually called psychokinesis (PK) – was also being investigated. One of the more careful experiments was the dice-throwing series conducted by Rhine and his collaborators. It consisted of a subject trying mentally to influence the fall of a set of dice. The dice were thrown by hand and by mechanical means. Strangely enough, the many successes in the first of a series of throws were followed by a rapid decline in success. It was found that the larger the number of dice being influenced – even up to ninety-six dice per throw – the greater the success.

In the early 1930s, the medium Rudi Schneider allowed himself to be investigated with scientific apparatus. Like all mediums of that era Schneider had to work in a dark room to be effective. This made it difficult to observe what he was doing, as well as to guard against fraud. These difficulties were resolved by using beams of infra-red light both as part of the apparatus and to detect anybody moving; this form of light did not disturb Schneider. When in his trance state he was able to cause reduction in the intensity of an infra-red beam in a suitably designed sealed box by up to 75 per cent. The absorptions of infra-red were discovered to be oscillating at about twice the medium's rate of breathing. The phenomenon depended on Schneider's presence but had no physical explanation at all.

The evidence for psychokinesis, telepathy and clairvoyance obtained in these experiments, though suggestive, was not entirely convincing. It was certainly not accepted by the majority of scientists: the results seemed impossible to reconcile with established science. They claimed that there were still some possibilities of error, either unintentional or intentional. It was unfortunate that automatic data-taking and the use of sophisticated equipment and analysis had not yet come fully into play.

With the advent of reliable and compact electronic equipment since the Second World War, deeper and more comprehensive tests of various forms of ESP have included work with animals. One of the first of these tests was reported in 1968 by two university scientists who, wishing to remain anonymous, used the pseudonyms Duval and Montredon. In their experiments the human element was almost entirely eliminated by the use of entirely automatic apparatus. The animals they used were mice, and their cage was divided into two halves, each of which could be electrified, to give a mild shock to the animal in it. The apparatus was designed to test whether the mouse could avoid the shock by guessing, ahead of time, which area would be electrified. A device known as a multivibrator was used to determine in a random fashion which of the two halves of the cage would receive the electric shock. The position of the mouse when the current was applied was established by two systems of light-sensitive cells, one for each half of the cage, the activation of which caused an electric pulse to be sent to an electric pulse counter.

Duval and Montredon were careful to avoid errors, such as the mouse being counted as being on both sides because it had jumped very rapidly, on receipt of the shock, from the electrified portion of the cage to the other side. Approximately 50 per cent of the trials were conducted at night, with no experimenters present; this eliminated the possibility of telepathy between the animal and its human investigators.

The initial results showed no significant effect: usually the mouse behaved mechanically; that is to say, it only jumped away from a particular side when it had been shocked. There were also mice who because of fear or fatigue just stayed in whichever half of the cage they had initially been placed. When the results from these particular animals were ignored there definitely seemed to be an effect resulting from precognition. Out of a total of 612 trials, the set of four mice tested avoided shocks fifty-three times more than might have been expected by chance alone. This success rate could only have occurred at random once in a thousand such cases.

Other groups of research workers have since repeated these experiments with additional precautions to ensure that the choice of which side of the cage to electrify was random, and to prevent the animal deducing any bias in the machine. The result they obtained was similar but here, out of 1,721 trials, the odds on getting it purely by chance were down to a hundred to one against.

To test if human beings can influence animals' movements researchers have tried to make protozoa (one-celled animals) move to a particular quadrant of a slide as seen in the field of vision of a microscope, though with little success. Attempts to influence the positions of gerbils and woodlice and other animals have done no better. However, the following animals seem to be decidedly more sensitive:

animal	*nature of capabilities*	*likelihood against chance*
mice	avoidance of future shocks	1 in 50,000
cockroaches	movements controlled by humans	1 in 8,000
chicks	turning on lights to keep warm	1 in 1,500
fertilized eggs	turning on lights to keep warm	1 in 1,800
cats	movements controlled by humans	1 in 60

Much more work has been done to investigate human powers since the introduction of automated apparatus. A mechanical dice-thrower has been used which only requires the experimenter to press a button for a throw to begin. After 170,000 throws one investigator found he had influenced the dice at a rate consistent with chance only once in a hundred times – significant, but not highly so.

Radioactive decay, occurring, for example, in the ultimate transformation of uranium to lead, takes place at a random rate. The decays can be used to generate a series of random numbers. These can be taken, for instance, as the times between each successive decay. Subjects have tried to influence the decays by altering the sequence of numbers so that it is no longer random but has order in it. Two schoolboys were able to make the changes with odds of one billion to one against this occurring by chance, whilst Helmut Schmidt of Duke University found that his subjects could achieve similar modifications with odds of ten million to one against chance. Other workers have since tried to influence this same process as well as other processes at the atomic level, but without the same success. Here we meet one of the snags of ESP research: the lack of repeatability. Thus one subject may apparently be able to affect some piece of apparatus whilst other subjects, working with other experimenters, cannot. Do we believe the successful result or the one that failed? It is not as clear-cut as in the physical sciences, where a strange result may be caused by new, experimental design or apparatus. In the realm of ESP, apparently contradictory results could also come from differences in the subjects selected for experiments: some may have powers, others may not. It is without doubt the difficulty of pinpointing precisely why repeat experiments are so rarely successful that has been the main cause of ESP research continuing for so long without its findings convincing the scientific establishment.

In some of the experiments described it is highly unlikely that the results have been obtained by chance. Yet most scientists have not taken them as *conclusive* evidence for ESP, partly because of the lack of repeatability mentioned above, and partly because the results are so contrary to modern scientific thinking. To continue just adding to the statistics of card guessing and similar tests will not bring about a rapprochement with science. Moreover, the cause of ESP is not helped when the explanations offered are mystical or scientifically incorrect. For example, it is downright wrong to suggest, as one worker in psychical phenomena recently has, that gravity may be the 'basic element of transition to be considered in connection with psychical phenomena', on the grounds that 'gravitation still represents a problem within physics, as it has not been proved to have an energetical aspect.' In fact gravity is the field of energy *par excellence.* All forms of energy are sources of gravitational field and gravity only arises from energy; gravity and energy are but two aspects of the same underlying substratum. Such specious theorizing will not get us anywhere! Only a wider range of experiments can help to track down all the physical aspects of the phenomena, and allow science to come to grips with ESP.

Russia has a great tradition of research into ESP, directed by competent scientists in well-equipped laboratories, bringing the physical nature of extrasensory powers into better focus. In the 1930s useful information on the nature of telepathy had been gathered by Professor Leonid Vasiliev, a distinguished physiologist, who pioneered these studies at the Bekhterev Brain Institute in Leningrad. Two female patients of a psychiatrist colleague proved extremely susceptible to commands to go to sleep given by a physiologist, Tomcshevsky. Both patients lay on a bed and the electrical resistance of their skin was monitored to detect a highly relaxed state. Each was required to keep pressing and releasing a hollow rubber ball linked to a pneumatic capsule, which traced a saw-tooth curve on paper tape. As soon as the subject fell asleep she stopped pressing the ball, so that the recorder would trace a straight line.

The first step of the experiment was for each of the women to be hypnotized by Tomoshevsky and, while under trance, ordered to fall asleep whenever he commanded it. Tomoshevsky then found it possible to cause the women to sleep, apparently when he sent out the command mentally. This was achieved with the hypnotist himself two rooms away while an observer stayed in the room with the previously hypnotized women. Nor did the observer know when the command to sleep would be transmitted; only the hypnotist knew that, and that only when a roulette wheel turned up a chosen number. After a period of sleep the hypnotist then commanded the women to wake. Even if one did not wake at the first attempt, several spikes appeared on the paper tape, indicating that the hypnotist's suggestion had been received.

Selective telepathy even proved possible. With three subjects and three hypnotists all in the same room each hypnotist was able to send one special subject into trance. It even transpired that when questioned by an observer each subject could correctly name the hypnotist who had commanded her to go to sleep.

The last phase of the experiments was to determine the physical basis of the influence transmitted from hypnotist to subject through thick brick walls. As Vasiliev said, 'We were all certain it could be nothing but radio waves. We could not imagine any other explanation. For we all know there are electric currents pulsating in the brain; and every alternating current sets up its electromagnetic field, or radio waves.' This idea had often been put forward, but on this occasion they decided to try it out in a further experiment.

What was needed was to screen out any possible electromagnetic waves that might be transmitted from the hypnotist to the subject. Two such screens were made, one of ironplate, the second of lead. The latter was essentially a lead capsule with a lead lid which slid along grooves containing mercury. In addition, the subject was placed inside an iron cage which helped to keep out electromagnetic radiation even more effectively. These elaborate shields failed to present any barrier. The subjects continued to go into a trance at the command of their own hypnotists, even maintaining the selective telepathy of the previous operation. Distance seemed to have no effect on the strength of the transmission; exactly the same success rate was achieved with a distance of 1,000 miles between hypnotist and subject. These results apparently destroyed the hypothesis that electromagnetic radiation was being used, since the lead and iron shielding were expected to deflect such radiation completely. This left nothing in its place. This hiatus was apparently so disturbing that the publication of the results was delayed till 1962 – Vasiliev admitted: 'We were dumbfounded. We were ourselves as if hypnotized by these unexpected results.'

The detailed evidence for this conclusion has not convinced some western ESP researchers. Even if the conclusion were accepted, one should consider very carefully the efficiency of the lead and iron shielding. There is a wide range of types of electromagnetic radiation, depending on its wavelength. This is the distance between the peaks of the radiation, like that between the waves on the sea. Light is that form of electromagnetic radiation with a wavelength of about five hundred thousandths of a centimetre; microwaves used in microwave ovens may have a wavelength of a few millimetres or so, whilst radio waves can have wavelengths as long as thousands of kilometres.

A room perfectly shielded against electromagnetic radiation of one particular wavelength might not be as impervious to a different wavelength. This is especially true with microwave radiation, which can leak through seemingly innocuous cracks. Structures round the walls of the room may become secondary sources of radiation because of resonance effects.

Since those Russian experiments, further tests have been made to monitor electrical activity on the scalp, thought to be related to underlying brain activity. A group of Russian scientists used a sender and receiver who had already shown some telepathic powers. For example, over a distance of 1,800 miles the mental image of a metal spring with seven tight spirals had been received as 'round, metallic, gleaming, indented, looks like a coil'; that of a screwdriver with a black plastic handle was recorded as 'long and thin, metal, plastic, black plastic'. In this experiment the receiver, Nikolaiev, was installed in a laboratory in Leningrad, where he relaxed into a state in which the predominant surface brain activity was of regular electrical oscillations of about 8 to 12 per second. These alpha waves, as they are called, occur in about 90 per cent of all humans on closing the eyes. The activity was made to change rapidly to the usual faster waking activity about three seconds after the sender had been told to transmit to Nikolaiev from four hundred miles away.

One of the research group summarized the nature of the changes in the brain as follows: 'We detected this unusual activation of the brain within one to five seconds after the beginning of the telepathic transmission. We always detected it a few seconds before Nikolaiev was consciously aware of receiving a telepathic message. At first, there is a general non-specific activation in the front and mid-sections of the brain. If Nikolaiev is going to get the telepathic message consciously, the brain activation quickly becomes specific and switches to the rear regions of the brain. This specific pattern remains clear on the graphs for some time after the transmission ends.'

When he was receiving images, the activity of Nikolaiev's brain was localized in the so-called visual area of the cortex at the back of the head, to which visual information is transmitted from the eyes. And when he was receiving sounds such as a series of buzzes or whistles telepathically, the maximum cortical activity was in the part of the brain called the auditory region, to which signals from the ears are relayed.

In their experiments in psychokinesis, the Russians have been lucky in the participation of subjects who appear to have the power to move objects without touching them. This dispenses with all the paraphernalia of innumerable guesses and the application of statistical methods. As the British parapsychologist Benson Herbert, who made a painstaking study of many of these psychokinetic phenomena, writes: 'But when K [a subject] slides some non-magnetic object ten centimetres observers need no knowledge of statistics to find this immediately convincing and dramatic. The problem is then simplified to an inquiry as to whether any normal means is employed.' Since there appears to be no shortage of subjects with such psychokinetic powers, these investigations have been going on for several years.

One of them, Nelya Kulagina, has been filmed in the process of moving objects at a distance. As seen on film, a compass needle, then the compass itself, was caused to rotate on a smooth table top when Kulagina's hands were held about six inches above them. She then caused matchsticks and a small non-magnetic metallic cylinder to roll off another table. Similar objects were made to shuttle from side to side inside a large plastic cube. She also apparently 'caused the yolk and white of an egg to be separated' when she was six feet away, and moved simultaneously five cigarettes placed on their ends under a glass jar.

When measurements were made of the electromagnetic field near Kulagina's body, the researcher, Dr G. Sergeyev, found that it was larger than that usual for humans by at least a factor of ten. Analysis of her brain activity during the session showed extreme activity in the visual area of her brain; even under normal conditions, the electrical potential at the back was fifty times greater than that at the front, whereas most people only have three or four times as much. During a psychokinesis session her heartbeat might increase to 240 beats a minute, four times its normal rate, and the magnetic field surrounding her body varied at the same frequency. There was even found to be a directional effect, in that she had focus in the direction of her gaze.

Kulagina has moved objects made of a range of materials: gold, aluminium, copper, steel, bronze, silver, glass, china, wood, paper. Even cigarette smoke contained in an upside-down glass bowl was cut in half by her efforts. She was also able to influence one pan of a pair of scales containing equal weights of 30 grammes, and held down one of the pans for 6–8 seconds. These efforts exhaust her as if she had used up a great deal of energy; during one of her sessions she was found to have lost nearly a kilogramme. The phenomena associated with Kulagina have been investigated by several groups of scientists, who could find no evidence of fraud such as her surreptitiously rocking the table on which the moved objects stood, nor of any explanation.

Another Russian woman with similar powers is Alla Vinogradova, who also has been filmed while causing objects to move. She can make round objects, such as a paper cylinder, roll on a flat surface and others slide. She holds her hand above the cylinder, a little to one side, and the cylinder starts to roll away from her hand; the force between her hand and the cylinder appears to be one of repulsion. She has made an aluminium cylinder swing round like a compass needle. With a metal chain round her wrist connected to an earthing point she is able to move an object weighing 100 grammes, while, if she is not earthed, she can only move a few grammes. But she appears to operate still better when the object she tries to move stands on a non-conducting table top. Electrical sparks may jump from her fingers to the object; the fluorescence of a neon lamp placed near her indicated electrostatic fields of up to ten thousand volts per centimetre around her.

These cases of psychokinesis probably have a far greater chance of being explained by some form of electromagnetic force than would the telepathic experiments; especially since there is already the evidence of extraordinary electrical and magnetic activity during these experiments on the movement of distant objects. But the evidence does not allow us to conclude that electromagnetism is the motive force. More experiments would be needed and unhappily they have not been done. In any case, in order to be sure we have as many pieces of the jigsaw puzzle as possible, we must bring in the even more recent phenomenon of the metal-bending. This may not involve a force of the same nature. It could provide pieces – but for a different jigsaw puzzle. We will have to keep the possibility in mind, but first we will look at more evidence on the phenomenon. It may well provide an important lead.

3 Experiments on Geller

Literally hundreds of 'tests' have been conducted on Uri Geller – on theatre stages, before television cameras, in aeroplanes, taxis, hotel rooms, restaurants and innumerable other places. The vast proportion have been carried out by newspaper reporters, television personalities, and friends and acquaintances of Geller. Very few have been held under what might be called scientific conditions, or with scientists directly involved. Even when scientists have expressed interest in carrying out experiments with Geller, as they have done recently in Britain, it has proved extremely difficult to reach him, surrounded as he is by a protective entourage of intermediaries whose interest may lie in values other than scientific ones. And naturally enough, the commercial interest arouses one's suspicion. In these circumstances it is a hard job for the scientist to obtain any data which will stand up to the scrutiny of his fellows. Since very few scientists have had access to Geller, the possibility of repeating experiments under carefully controlled conditions appears remote.

In fact, there have been two series of tests on Geller, which go part of the way towards satisfying the strict criteria needed to dispel our suspicions of fraud. The experimental conditions have been specified in enough detail to allow other researchers to duplicate them. But during these early experiments it did not prove possible to monitor and describe Geller's psychological and physiological state. This is generally a difficult problem – how to delineate this state for *any* person is still not clear.

Both series of experiments were conducted at the Stanford Research Institute, California, under the direction of Harold

Puthoff and Russell Targ, two physicists trained in lasar research. Most of the sessions were filmed or videotaped, so that any kind of mass hypnosis which might have occurred could be detected; no materials used in the tests were supplied by Geller. Some of the experiments were scrutinized by a magician on television monitors for possible sleight of hand procedures (none were in fact detected); the experiments were performed on what is termed a 'double-blind' basis wherever possible, which means that neither Geller nor the experimenter could know the correct answer beforehand.

The first set of experiments took place over a period of six weeks in November and December of 1972, involving a variety of tests, some with astoundingly successful results. One of these was dice-guessing. A single dice was placed inside a closed metal box, which was then vigorously shaken by one of the experimenters and placed on a table. Nobody could see how the dice had fallen. Geller then tried to guess which face of the dice was upwards. He made eight guesses – and every time he was right. But he would not hazard two further tries as he did not feel confident that his perception was clear enough.

If we exclude these two refusals as true attempts, then the odds in favour of his guessing by chance the upper face of the dice eight times in succession are about one in a million. In assessing the probability, it seems justifiable not to take the other two throws into account since they were not even attempted. Only if Geller had tried *and failed* in these two attempts would they have to be counted.

The next test was to guess in which of ten identical aluminium film cans was hidden a certain object. After an assistant had put the object in one of them, the cans were placed in random positions on a table. The assistant then left the experimental area. Geller then entered the room with experimenters, who equally had no idea which can contained the object. Geller either passed his hands over the cans or simply looked at them. He then indicated which cans he thought were empty, and these were removed; when only two or three cans remained Geller stated which can he thought contained the object.

In this experiment Geller was successful each of the twelve times he made a guess. As in the previous dice-guessing test, he declined to guess on two occasions, having difficulty once with a metal ball-bearing wrapped in paper and once with a sugar cube. Both of these 'failures' can be neglected for the same reason as before. He was successful in finding water (twice), steel ball-bearings (five times) and small magnets (five times). The odds of his twelve correct guesses being due to chance were again one in a million, a small number indeed.

Two experiments were next performed to test Geller's powers of psychokinesis. In the first of these, a precision balance had a one-gramme weight placed on its pan before being covered with a glass jar. The balance generated an electrical output voltage proportional to the force applied to it, and the voltage was recorded on a chart in the form of a continuous strip. On several occasions Geller made the balance move, and a displacement was recorded on the strip chart. The movements were ten to a hundred times larger than could be produced by striking the table on which the balance stood, hitting the glass case covering the balance pan, or by jumping on the floor (none of which Geller, under close observation throughout, could have done). On one occasion the deflection on the chart was equivalent to a decrease in weight of the balance pan of about one and a half grammes; on another, to an increase in its weight of nearly a gramme; each happening for about one-fifth of a second.

The second main psychokinetic experiment was that of seeing if Geller had a magnetic field. He passed his hands near the probe of a Bell gaussmeter, an instrument which measures such a field. Geller was able to cause a full-scale deflection of the instrument a number of times (recorded on film) showing that he had an effective magnetic field at least half as strong as the earth's. This effect may not, in fact, have been due to any magnetic field he himself possessed, but to some direct interaction he achieved with the measuring apparatus, such as affecting its electronic circuitry. The experimenters are said to have also watched the movement of iron filings lying on a sheet of paper when Geller's hands were nearby; this does seem to mean that there was perhaps some magnetic field effect caused by Geller's presence.

He briefly tried his hand at metal-bending, but because of his need to touch the metal the effect was not taken seriously. Nonetheless, the actual force required to bend some metal rings to the extent that Geller had distorted them was found to be the considerable amount of 150 pounds.

The remaining tests were all devoted to telepathy. Seven simple pictures were drawn on filing cards before Geller arrived at Stanford, and each was sealed in an envelope by an outside assistant. At the start of the experiment an envelope was selected, opened by the experimenters and the picture in it identified. The scientists then went into the experimental room and asked Geller to guess what was the drawing inside the envelope. He guessed with almost complete accuracy in each of the seven cases.

Even more dramatic was Geller's performance in the second set of experiments at Stanford Research Institute, lasting eight days in August, 1973. The drawings and their senders were all some distance from Geller during these tests, so that the possibility of his receiving any clues was zero. Furthermore, Geller was either kept in an electrically shielded room or else the pictures he was trying to guess were drawn on the East Coast, three thousand miles away, thereby making the surreptitious use of something like radio signal apparatus almost impossible.

The drawings in the first experiment were obtained by choosing at random from a dictionary any noun that could be drawn. This was done after Geller had gone into the shielded room, the sender and drawing staying always outside. The first word chosen was 'fuse' and the subject drawn was a firecracker. Geller said he saw 'a cylinder with a noise coming out of it' and his drawing looked like a drum with a number of other cylindrical

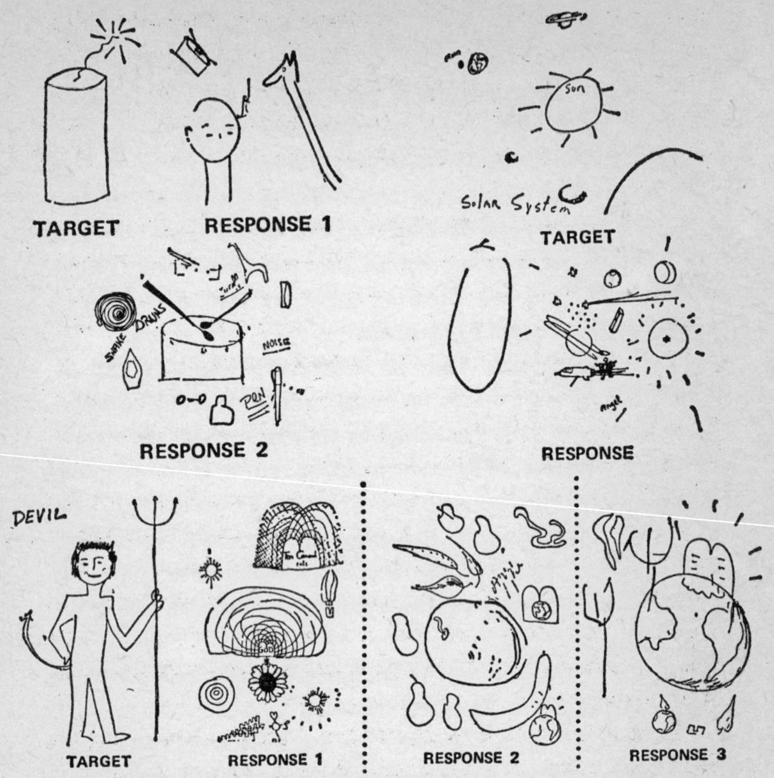

Target pictures and responses drawn by Uri Geller under scientific conditions, as first published in the scientific magazine *Nature* in 1974 (Stanford Research Institute, courtesy Doctors Puthoff and Targ)

objects. The second word chosen was 'bunch', and so a bunch of grapes was drawn. Geller said he saw 'drops of water coming out of the picture', then mentioned 'purple circles'. He then drew a bunch of grapes, exactly the same in number, twenty-four, as in the original drawing and of nearly identical shape.

Geller and Dr Puthoff were then locked up in a shielded room together, half a mile away from the experimenter's office. The next drawing was of a devil with a trident; Geller found this difficult, and produced three different drawings, composed of

Moses' tablets with the Ten Commandments inside the earth, with a trident outside; of an apple with a worm coming out of it (with a snake also in the picture); and finally one which was a mixture of the previous two with God inside. The difficulty Geller had experienced in drawing the devil may well have been cultural.

One of the experimenters was then incarcerated in the shielded room, while Geller stayed outside with the other experimenter. The picture Geller had to guess was of a solar system; he drew a picture which was a very close resemblance, with Saturn and its rings and a central glowing sun. The next task was to guess, while outside the shielded room, a picture placed inside it drawn earlier by a scientist who did not belong to the experimental group nor was present during the test. The power being tested here was that of clairvoyance, of being able to 'see' what would be to normal senses a completely invisible object; the lack of involvement of the person who had drawn the picture ruled out the possibility of telepathy – i.e. of any thought transference. Geller was unable to succeed. Two further tests were tried with Geller's brain waves being monitored whilst he was in the shielded room, and he failed again.

His next two attempts were only moderately successful, but the third was totally so. The original drawing was of a seagull in flight; he said he saw a swan flying over a hill and drew several birds, one almost identical with the original. He was also very close to the picture of a kite drawn by a computer on a television screen; he was about 150 feet away from the original, which was in a different room. He was relatively successful in guessing the picture of an arrow through a heart stored in the computer memory, though he failed in another such case. This was not a clear-cut test of clairvoyance, however, since there were several people in the computer room who knew the nature of the stored target. There were two attempts at long-distance communication, one resulting in some similarity, but the other not so close to the mark. To summarize, out of fifteen drawings which

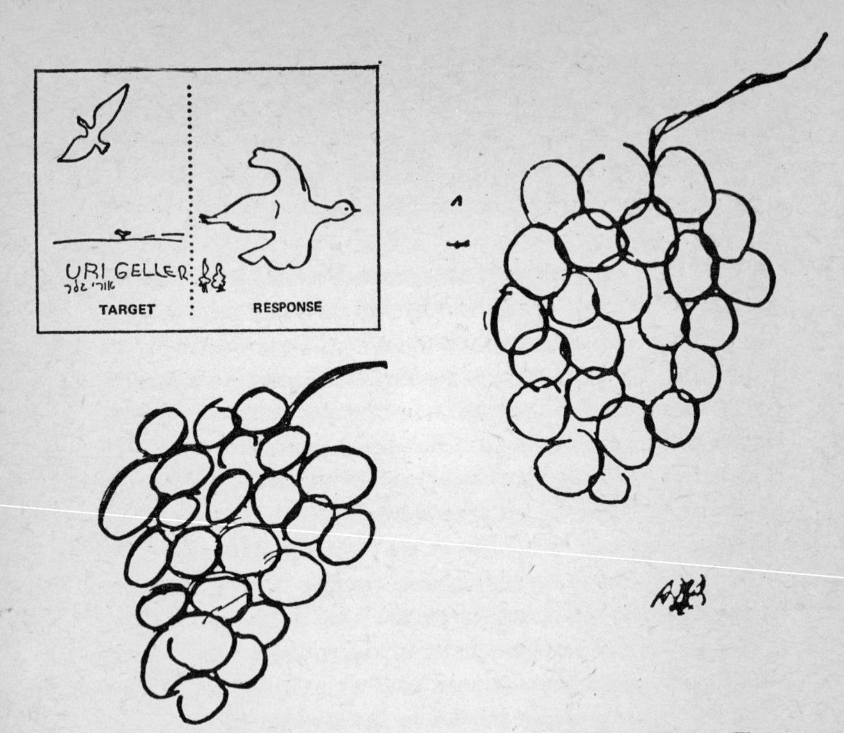

TARGET **RESPONSE**

Two successful responses by Geller during the Stanford Research Institute series of tests. The accuracy of the drawing of the grapes is astounding
(Stanford Research Institute, courtesy Doctors Puthoff and Targ)

Geller tried to guess, he was undeniably successful in seven of them, to some extent in four others, and made no attempt with the remaining four. This is not a bad score. We should expect that the mere chance of guessing what had been drawn by the experimenters in most of these cases would be almost zero.

The bunch of twenty-four grapes is here a very good example. One way of assessing this is to consider the original drawing as consisting of a set of seven rows of grapes, containing 1, 3, 3, 4, 6, 4 and 3 grapes respectively (starting at the bottom). Geller's bunch of grapes has corresponding numbers 1, 3, 4, 4, 5, 4, 3, the third row having one too many, the fifth row one too few. If we assume that these seven figures were guessed at random, each being chosen from the digits 0 to 9, the odds for Geller's choice being identical with the original drawing in five of the numbers by chance are one in one hundred thousand.

This calculation overlooks the fact that the other two numbers, which were wrong, were also very close. Nor does it take into account that Geller guessed the number of rows correctly; this introduces another factor of at least ten in the odds against chance (since it is unlikely that there would be more than about ten rows in a bunch of grapes). But then it is necessary to include the chance of it being a bunch of grapes in the first place – 'purple circles' as Geller described them. There could have been at least a hundred objects to choose from, if not more, and that they had to be in a bunch and not in a line or other shape is also improbable by a factor of at least ten. Taken all in all, the odds for Geller guessing the details of this picture by chance alone are at least as low as one in a thousand million, and this is definitely an underestimation: a remarkable result, and one only probable in terms of some form of communication.

The major portion of the research with Geller on remote perception graphic material appeared in *Nature*, one of the world's most respected scientific journals, in October 1974, along with similar results on other subjects. There was an accompanying editorial comment which criticized the report as 'weak in design and presentation', giving 'uncomfortably vague' details of various safeguards and precautions introduced against the possibility of conscious or unconscious fraud on the part of one or other of the subjects, and 'not concentrating in detail and with meticulous care on one particular approach to extrasensory phenomena'. At best, it was concluded, the tests were more 'a series of pilot studies ... than a report of a completed experiment'. In spite of these considerable criticisms the paper *was* published by *Nature*, partly to put the results 'in more reasonable perspective' than the considerable advance publicity attending the paper had given it, and also to allow other scientists to gauge the quality of the particular research.

Criticism was much heavier from other areas, highlighting supposed inadequacies in the experimental procedures which could have allowed deception to occur. Some of these criticisms have since been found to be invalid. In particular, the supposition that some sort of electromagnetic device was secreted by Geller in his teeth or elsewhere in his body has been ruled out, because it would have required collusion on the part of some of the experimenters, as well as being impossible to achieve by present technology. Nor could the dice used have contained small radio transmitters signalling which side was up, unless they had been put there by the experimenters themselves.

Targ and Puthoff themselves summed up their results as follows: 'As a result of Geller's success in this experimental period, we consider that he has demonstrated his paranormal ability in a convincing and unambiguous manner.' From the calculation made above for the bunch of grapes, this conclusion would seem to be fully justified. There appears to be no chance of fraud in the experiment, unless there was gross collusion between all concerned. If such a charge is to be made, then there must be good backing for it. From other evidence presented in the first chapter, together with the filmed evidence described earlier, there seem no grounds at all for suspecting fraudulent behaviour. Nor could the collusion have only been on the part of the three principal characters; other people were involved in the various experiments. This, allied with the disinterested character of the investigation, renders any such charges unfounded and we need not consider them further.

The results obtained at the Stanford Research Institute are in some ways very difficult to accept. Nearly every scientist of my acquaintance has reacted by saying that he will believe only if he sees with his own eyes. Even then many scientists would find the evidence still hard to accept. To bring about any rapprochement between their scientific knowledge and what they had observed would, they say, need further experiments aimed at discovering as much as possible about the nature of the phenomena. Such tests, it is hoped, would indicate how effects of this sort might be consistently obtained so that they could then be investigated by any scientist desiring to do so. Then, in this way, general acceptance of the findings and, one can only hope, the understanding of them will come.

The phenomena of guessing pictures or objects and of metal-bending or affecting instruments have various important features in common. In every situation at least two physical objects are involved: the subject's brain which he uses to achieve the effect in question, and the object with which the subject communicates, be it the brain of the experimenter who created the drawing he is trying to guess, the actual drawing itself, the piece of metal the subject is trying to distort, or the instrument he is attempting to disturb. What additional agents are present is unclear; there may, as many suggest, be elements of the subject's brain activity which are not physical and therefore not measurable, on the one hand; and on the other, there could be some external 'cosmic' influence.

A scientific investigation can only look into that which is physical. The non-physical realm, be it 'mental', spiritual, 'aetheric' or any other, so far has defied scientific analysis. There are certainly ways of obtaining knowledge about the non-physical aspects of experience which provide helpful insights into their nature, but the results so obtained are difficult to describe and to relate in detail to the general corpus of knowledge. They are almost impossible to correlate with our picture of the material world constructed hitherto so effectively by means of the scientific method. However, it may be worth the struggle to understand extrasensory phenomena in scientific terms because this may in turn allow us to use the knowledge to open up valuable new directions in science. And the Geller phenomena could perhaps be a stepping stone.

But we may fail in this quest, and we shall not be the first to do so. Many great scientists have tried: Sir William Crookes, Sir Oliver Lodge, Lord Rayleigh and many, many others; and all failed. At least we can take heart at the fact that we are starting without the bias which affected some of those earlier researchers.

Indeed, nowadays a scientist must take care to avoid being so biased in the other direction that he is over-sceptical. He may well refuse to believe what is staring him in the face because it cannot be fitted into his materialist view of the universe. There may already be hints on how to unravel the mysteries of ESP in the vast literature on the subject. Yet a scientist may neglect them because they strike him as unreliable. They could, however, just be poorly presented by their originators and so be unfairly consigned to scientific oblivion. Winnowing the wheat from the chaff must be done with care.

For the scientist, it is also necessary to prepare for many shocks to one's scientifically trained system during the endeavour. One clear observation of Geller in action had an overpowering effect on me. I felt as if the whole framework with which I viewed the world had suddenly been destroyed. I seemed very naked and vulnerable, surrounded by a hostile, incomprehensible universe. It was many days before I was able to come to terms with this sensation. Some of my colleagues have even declined to face up to the problem by refusing to attend the demonstrations of such strange phenomena. That is a perfectly understandable position, but one which does not augur well for the future of science. I myself am still fighting hard to preserve the scientific understanding of life as the legitimate one. For me, the attempt to explain the Geller phenomenon is foremost in this. The attempt may fail, but now the stakes are higher than they were. Science has evolved so much since a century ago. It *should* be able to give a satisfactory answer. If not, the scientific method truly will have been found wanting, and could well suffer a blow from which it might never recover.

If ESP is fully authenticated, one has to accept it – truth will out. But if scientific explanations fail, what then? This is a question we can leave, however, till it cannot be evaded. The priority is to find out if science can possibly give the facts and mechanism – the *what* and the *how* – of the Geller phenomenon. Only when it has decisively failed to do so must the scientific approach be jettisoned.

The features of the Geller phenomenon to investigate scientifically are the two physical ends – Geller's brain and the inanimate object or the brain he is in contact with. Brain activity is notoriously difficult to monitor. It is very complex and the crucial variables describing its activity are not known. What is more, workers at the Stanford Research Institute had difficulty with Geller in that he did not relish having electrodes attached to his scalp to monitor his brain waves. Even with complete co-operation on Geller's part, the fact is that the brain still presents an enormous puzzle to science. Until efficient apparatus is designed which can scan activity at various levels of the brain by remote means, it is perhaps unlikely that it will yield up its mysteries – the nature of will, intellect and emotion. That day may not be very far off, but it certainly is not going to help now in solving the problem of the Geller phenomenon. We are therefore left with the other end, the object on which Geller and others like him act. It makes sense to confine our investigations to what might be called 'the Geller effect' – that of metal-bending pure and simple. Because they are as yet too elusive to be scientifically analysed, one must omit telepathy, clairvoyance, and all the other phenomena of ESP from present studies. Not that useful and important work cannot be done even now, but experiments designed to uncover possible physical mechanisms for these phenomena may well be difficult to come by. The greatest hope lies therefore in the Geller effect.

The scientific tests which Geller was subjected to at Stanford did not probe deeply into metal-bending, owing to the possibility that he might have been causing it by physical pressure. To avoid any likelihood of this it is necessary to do one of two things. On the one hand, tests must be carried out in which such force as is applied during the bending is measured while the experiment is in progress. If less force were used by Geller during a bending session than would be needed to cause the metal to bend by the amount it actually had, then some unknown, non-mechanical force would have been responsible and the Geller effect authenticated. This approach is at present under way, but requires delicate measuring devices embedded in pieces of metal to register the amount of force actually applied.

A more direct test, which precludes the question of the mechanical force applied during bending, is to require that metal be bent without Geller actually touching it. If he can achieve that, then the Geller effect is indeed beyond current scientific understanding. Geller did not achieve any well-authenticated bending at a distance during his time at Stanford, but on 2 February 1974, during one of his visits to England, a successful test of this kind was carried out with Geller by myself. Pieces of metal (aluminium and copper), strips of various types of plastic, some single crystals of potassium bromide which were long enough to be stroked like pieces of cutlery, various wire mesh tubes and a sealed glass tube containing a strip of aluminium – all these were used to try out Geller's powers. In addition, a small but sensitive Geiger counter – to detect radioactivity — and a primitive detector of ultra-violet radiation were included in the apparatus.

The various strips of metal and plastic and the sealed glass tube were laid a few inches apart on a metal sheet. A strip of aluminium, placed inside a wire mesh tube with its end firmly fixed, was also laid out. The objects had been prepared in the metallurgical department of King's College, London, and there was no chance of Geller having been in contact with them before the experiment. Two of my colleagues were also in the room with Geller, acting as observers.

The various strips of metal and plastic were carefully scrutinized at the beginning of the test to confirm that they were straight. First of all, Geller tried to bend a metal rod without touching it, but he did not succeed. It was then observed that one of the aluminium strips lying on the tray was now bent, without, as far as could be seen, having been touched either by Geller or by anyone else in the room.

To see whether Geller could repeat the metal-bending feat of the Dimbleby programme, he was then handed a teaspoon which had been brought along with the other materials. I held the bowl end while Geller stroked it gently with one hand. After about twenty seconds the thinnest part of the stem suddenly became soft for a length of approximately half a centimetre and then the spoon broke in two. The ends very rapidly hardened up again – in less than a second. There was also, as far as could be determined by touch, a complete absence of heat at the fracture. This sequence of sudden softening and complete loss of cohesive strength, breakage, and then rapid hardening was almost identical to that observed when the fork broke during the Dimbleby programme. Here, under laboratory conditions, we had been able to repeat this remarkable experiment. Geller could simply not have surreptitiously applied enough pressure to have brought this about, not to mention the pre-breakage softening of the metal. Nor could the teaspoon have been tampered with – it had been in my own possession for the past year.

Then Geller gently stroked a single crystal of potassium bromide about two centimetres long, and it split into two pieces within ten seconds. It was difficult to assess the force that had actually been applied to the crystal, but subsequent tests have shown that such crystals cannot be broken by gentle stroking alone. To show, of course, that pressure was not the cause of the crystals breaking and that this was a paranormal effect, it would be necessary to measure how much pressure Geller had in fact applied. Geller also stroked a thin wooden strip with no result. When he held his hands above a blue plastic strip this became discoloured. Such discoloration is normal when bending these plastic strips, though Geller was not in fact able to bend it without touching it.

After this series of tests the objects on the tray were re-examined. It was found that the last five centimetres at one end of the aluminium strip in the closed wire mesh tube was now bent with a radius of curvature of about five centimetres. One should bear in mind that Geller was continually under the scrutiny of the two observers. He could not unseen have opened the sealed tube containing the aluminium strip and interfered with it. Indeed, he was occupied trying to bend other objects at that time. Furthermore, there was no evidence of any tampering with the end of the tube.

At this point the primitive ultra-violet detector was used. This was constructed of a strip of annealed aluminium, covered with a thin layer of sodium salicylate, and sealed in a partial vacuum within a quartz glass tube. In darkness the salicylate layer glows purple in response to ultra-violet radiation. This can also be produced by rubbing the tube so as to generate a high static electrical potential on its surface by means of friction; electron emission will then produce ultra-violet radiation. Geller attempted to bend the metal inside the quartz tube, first of all without touching it and finally, when that failed, by stroking it gently. The purple fluorescence he produced was no greater than what might have been expected from the static electricity generated by friction. As the metal strip in this tube did not bend, the result, although it did not give any support for ultra-violet radiation as the agent causing bending, equally could still not allow us to rule it out completely.

The final test was to determine if Geller could produce a deflection on the Geiger counter; this should indicate whether he could produce radioactive radiation. When it was held near him, Geller registered a zero count on the instrument, taking into account the average background rate of about two counts per second produced by cosmic rays coming from outer space. Geller then took the monitor in his own hands and tried to influence the counting rate. We all stood round looking at the dial and listening for the tell-tale tone.

At first nothing happened, but by extreme concentration and an increase in muscular tension associated with a rising pulse rate, the needle deflected to fifty counts per second for a full two seconds, the sound effects heightening the drama of the occasion. By means of a small loudspeaker each count produced a 'pip', and before Geller affected the machine the sound was of a steady 'pip . . . pip . . . pip . . .'. In his hands the sound suddenly rose to become a wail, one which usually indicates dangerous radioactive material nearby. When Geller stopped concentrating the wail stopped and the apparent danger with it. This wail was repeated twice more, and then when a deflection of one hundred counts per second was achieved, the wail rose almost to a scream. Between each of these attempts there was an interval of about a minute. A final attempt made the needle deflect to a reading of one thousand counts per second, again lasting for about twelve seconds. This was five hundred times the background rate – the machine was emitting a scream in the process. After a rest of several minutes, a further deflection of two hundred counts per second was produced, lasting about five seconds.

At the end of the session the Geiger counter was tested to see if its counting rate could be modified by pressure on the monitor to produce the same effect. Despite the considerable force applied, no change came in the counting rate from that caused by background radiation. It therefore seemed unlikely that Geller had achieved this effect by distortion of the monitor head.

The conclusions of these experiments were threefold. First of all there was clear authentication of the metal-bending effect, both by the distortion of the aluminium strip in the closed wire-mesh tube and by the rapid breaking of the teaspoon. Secondly, that metal could be bent inside the mesh enclosure indicated that if electromagnetic radiation were causing the effect at all, then only a restricted range of wavelength could have been responsible. As mentioned in the previous chapter, the various forms of this radiation are differentiated by the size of their wavelength – the distance from one peak to the next. Radiation with wavelength longer than the holes in the mesh would have great difficulty in penetrating the tube; it would not be able to wriggle its way in as radiation with much shorter wavelength could. Only radiation with wavelength of less than a centimetre would be able to penetrate this type of tube and allow the bending of the contents.

The third conclusion to be drawn from the experiment was that a wide range of effects can occur, including causing a Geiger counter reading to be considerably modified. We therefore had to go one step further and investigate the variety and possible cause of these effects. We have some anecdotal evidence on the variety, such as that Geller can influence computers to make them malfunction or even stop working completely. Ed Mitchell, the ex-astronaut who sponsored the Geller experiments at the Stanford Research Institute, found that the pocket computer he was carrying sometimes failed to function in Geller's presence. It has even been said that Geller can disturb any scientific instrument if he so wishes. He certainly could modify a Geiger counter, and he also affected a Schmidt generator, a machine which is controlled by radioactive decay, an atomic process which will be explained later. This instrument has nine bulbs arranged around the circumference of a circle, and the light moves either clockwise or anticlockwise at random from the lit bulb to the next one, as determined by the radioactive decay. According to Dr Ted Bastin, who performed this test on Geller in February 1974, Geller was able to cause the movement of the light to proceed according to his own dictates.

At this point it is clear that the metal-bending has led to a very complex set of new phenomena. The evidence points to effects which may be occurring at the atomic level, especially those involving the Geiger counter and the Schmidt generator. On the other hand, there are also the observations we made of the metal becoming plastic, an effect which occurs over much larger sizes than the atom. There may well be here more than one process at work.

This may not only be true at the object end of the metal-bending process, but also at the subject end. Geller's physiological state while causing the Geiger counter deflection appeared very different from that while metal-bending. Unfortunately, he had not been wired up to discover exactly what his heart and brain activity was in each case, but there undoubtedly seemed a distinct difference. His extreme physical tension while acting on the Geiger counter parallels very closely that of the Russian subject, Kulagina.

The only way to find out exactly how complex is the Geller phenomenon is to experiment further. And by great good fortune the wide interest first generated by the Dimbleby programme has given us a number of other possible subjects, not only with the same powers, but eager and willing to offer their services.

4 Other metal-benders come forward

Geller's three consecutive British appearances, on the Jimmy Young, Dimbleby and Blue Peter programmes, set off a rash of spoon-bending all over England. Because of today's rapid communications, the reaction was greater and swifter than in the table-rapping days of the Fox sisters in America back in 1849. Scores of people suddenly discovered themselves to be metal-benders and hundreds claimed they could start or stop watches at will. This initial burst of metal-bending was followed by a steady increase which still continues. People who think they can 'Geller-ize' cutlery are coming forward at the average rate of one or two a week. And there have been further spoon-bending mini-epidemics whenever Geller reappeared on British radio or television. Radio and television telephone switchboards have been flooded with calls reporting Geller phenomena for hours after such programmes. Geller's appearance on the Jimmy Young show on 29 October 1974 produced such an effect, with hundreds of people telephoning in. One such case had its amusing side, when a policeman just about to go on duty found that his hat badge had started to bend. As he commented ruefully, 'I'm not sure how I'm going to explain *that* when I get round to the station.'

It has even been possible to devise a sort of scale in metal-bending ability: grade 1 (the lowest), where the claimant functions only to the accompaniment of Geller either on radio or television; grade 2, where the power appears as a direct after-effect of a Geller performance; and grade 3, where the claimant appears to function quite independently.

In grade 1 we have a preponderance of adult strokers, among them a middle-aged business man living in the South-east. He was listening to a radio discussion about Uri Geller on the Jimmy Young Programme and at the same time eating his lunch. 'As is now my habit whenever any programme concerning Geller is on,' he says, 'I was gently rubbing a dessert spoon lying on the table. Not only did this spoon bend at right angles, but two other kitchen implements also bent . . . In fact [my stainless-steel chip pan] looks as if a giant has screwed it up, and another kitchen gadget has been folded in two.' He notched up a further score when he got on the telephone to Jimmy Young about what had happened, causing a key near him to bend while they were speaking of Geller.

In grade 2, where children share the honours with adults, we have a lady in Wales whose husband witnessed her success in bending a fork, the prongs of which he was holding. Her method had been to stroke it gently at the neck. So far she has not been able to repeat her performance. The Geller impetus seems to be vital in these cases, although in this one it has at least lasted for a while after his disappearance from the television screen. And what is one to make of the several Swedish women who have attributed their pregnancies to the fact that, as a result of a Geller programme transmitted on video-tape, their contraceptive loops had become too distorted to function efficiently? Geller was not himself in Sweden at the time, but the effect of his recorded programme on the cutlery of scores of Swedish homes was as dramatic as that during the programmes which had previously gone out live in Great Britain.

The question we have to ask ourselves now concerns the nature of the causal chain setting off these manifcstations. One can feel fairly certain that Geller acted as a catalyst in helping people discover powers they had not suspected in themselves. But the sad sequels to the video-taped programmes do seem to put paid to the notion that Geller can be regarded as the source of some type of direct radiation responsible for causing them.

This brings us to grade 3 (most powerful) manifestations, where the parties seem to be able to bend metal quite independently of Geller's appearances. In this group it is the children who predominate. Among them is a boy who was asked along with his brothers if he could bend metal. He hadn't seen or heard a single Geller programme and had to be instructed how the gentle stroking should be done. After half an hour, the spoon he was stroking acquired a very visible deflection.

Our next case involves a young lady of twenty-eight. She was asked by a member of the Magic Circle, while they were sitting together in a restaurant several days after the Geller programme, to try to Gellerize a spoon. The magician later wrote of the experience: 'After three or four minutes it started to bend. Finally it moved right over. [She] left the spoon in her car overnight and it was still bent in the morning. Last Saturday I spoke to her on the phone . . . and she told me that while she was holding the spoon in her hand the top of it then actually fell off . . .' Both cases occurred quite independently of Geller.

There is a broad age-range for people who possess the power of metal-bending, but a clear pattern emerges from the cases that have so far come to light. Of those nine with the weakest powers (needing Geller's image), eight are adults and one is a ten-year-old girl; all those eight adults are married women. There are only two cases with grade 2 metal-bending powers (Geller's image present not too long before testing), and again they are both married women.

Our final or grade 3 category of people with strong metal-bending powers contains thirty-eight, of whom only one is a man, three are women and all the rest are under seventeen, fourteen being boys and twenty girls, and of them seven are severely retarded. The lowest age among those with evident spoon-bending powers is seven, though a case of a four and a half-year-old girl who can bend spoons has been reported, but not yet fully authenticated.

It is interesting to note the almost complete absence of adult males on this list; excluding Uri Geller only *one* case is known of a male over twenty years old possessing this power. Against this, there is the high proportion of children. Why? Does the developing brain apparently possess some property endowing it with the ability to bend metal until man's estate is reached, only to wither away?

One might begin to answer this by considering the possibility that children may have spent more time in experimenting with metal-bending than adults. There is little detailed evidence on this, but from the hundreds of letters I have received from people who claim to have some form of power, about as many adults as children appear to have tried to bend cutlery at some time or other.

As the next possibility one could conjecture that metal-bending is a power which may have survival value in the young but is not needed in the adult. This seems unlikely, and more probable is the idea that the ability to bend metal may be a secondary consequence of the ability to communicate telepathically, assuming a common mechanism. After all, telepathic ability may not be so unrelated to metal-bending, since both involve effects at a distance.

In any case we must treat the figures with caution and be guarded in our speculations. So far, we are concerned with a total of forty-six people – that is, roughly one in a million of the British population. There are others who still require further investigation, not, as yet, made. At most, however, there are no more than five hundred who have come forward with any claim at all to these strange powers – only one person in a hundred thousand. There may be many more who have the ability but have not yet realized it. Indeed, everybody may have the power to some extent. But although there must have been millions of children in Britain who tried to bend cutlery during or after the Blue Peter programme, only forty or so presented themselves.

These provisional figures on metal-benders are still striking enough to pose a problem. Naturally there have been strong reactions to the emergence of the strange power of spoon-bending. Some parents of children with the ability at first thought their children were bending the cutlery by force when they were not being watched; the parents promptly bent it back into shape. The metal-benders' school teachers have run through the whole range of reactions, from the sympathetic to the markedly hostile. Science teachers tend towards the latter, some even refusing to watch any demonstration of the child's ability. One head teacher is reported to have said, 'I thought only God could do that.' Fellow pupils are more open-minded and are prepared to believe what they see. The metal-bending children themselves take the possession of their powers quite casually: much in the world about them is inexplicable, so why should they worry about metal-bending?

It has not been so easy for the adults with these powers to come to terms with them. The reactions of colleagues are far more difficult to live with. The only man among the most powerful metal-benders has talked to hundreds of acquaintances, and has found that the majority treat him with suspicion – as if they were thinking, 'He's all right really, I suppose, but what a pity he pretends he can bend cutlery.' What they actually come out with is a phrase I have heard often from the lips of my own associates: 'Hide the spoons!'

Later reaction to the announcement of metal-bending powers has in many instances been one of fear. The dread that the power may be evil has caused some parents to try to stop their offspring's metal-bending activities. In a recent case this proved impossible; cutlery and other thin metal objects in the house were getting bent at times when the child was not even attempting it. The parents felt very persecuted – especially when neighbours suggested these powers might 'come from the devil'.

There is no doubt that if people really can cause metal objects a distance away from them to bend, then they could be capable of causing great harm. One has only to consider the terrible consequences of an aircraft or other engine being crippled by the bending of one of its parts. Geller has, in fact, already claimed to have stopped a ship by damaging its fuel supply system. In modern ships controlled by sophisticated electronic equipment there is an even greater risk of damage.

Any capacity by which energy can be rapidly liberated can be harmful – an ordinary blow with the fist, or a bullet from a gun, for example. Even a car can be misused with destructive effect. It is up to the individual to ensure that when he handles sources of power he does not use them for evil ends. Complicated issues may be involved, but the principle at least is clear: never set out deliberately to do harm; on the contrary, always avoid it. Control is, of course, the key factor, and in the case of metal-bending, there is no real control of the power. It seems to flash on and off when least expected. A seven-year-old girl may be sitting dreamily watching television and the piece of metal she is gently stroking will suddenly bend. It may even do so at such a speed that it will trap her finger, and she will painfully have to extricate it. That control of the power which appears to be almost entirely lacking will remain so until the phenomenon is better understood. But anyone suspecting metal-bending powers as being inherently evil may be reassured. From direct observation I can say that there is absolutely no evidence of the ability in itself being evil. That it resides mainly in children, some very young, and all of apparently stable personality (excluding the retarded ones), strongly encourages this conclusion.

That is not to say that the ability to bend metal carries no danger either to the children themselves, or to those around them. Whatever force field is in operation causing cutlery or other metal objects to bend may well affect tissue, blood or bone. There are already reports that some of these children possess the power to heal with the laying on of hands. On the other side of the ledger, headaches and similar ailments are alleged to have increased in the families of metal-benders. There have also been cases reported of children falling more frequently into trance-like states as their metal-bending prowess has developed. They can still function automatically in such states, and are therefore liable to be exposed to danger.

At this stage too little is known properly to assess the dangers associated with metal-bending, although in one case already I have had to advise the parents to put a stop to it; both the son and the daughter appeared to be getting out of control. On the other hand, I have had the brain waves of children monitored while they were bending metal, with no abnormality appearing. However, until the phenomenon is better understood and the real dangers are known, the greatest care on behalf of the physical well-being of all concerned must be exercised in each and every case.

The first query to be raised about the metal-bending abilities of the hundred or so people who have so far come forward is: is their ability genuine? To many people, and to scientists especially, the phenomena reported by these metal-benders and people observing them are impossible to accept unless mechanical force can be shown to have been applied. A typical response came from Dr Christopher Evans, a psychologist. He bent a fork backwards and forwards several times until it was ready to break. He then took it into his laboratory canteen and proceeded to 'Gellerize' it, thus pointing out how easy it is to take people in.

The various accounts of Geller's powers in the first chapter give very strong evidence in support of their authenticity. Metal objects were bent when held by others or even when they were not being touched. Similar experiences where objects have been bent at a distance have been observed among the newly discovered metal-benders.

As in Geller's case, most accounts of these other metal-benders have come from newspaper reporters. One, about an eleven-year-old girl, stated: 'Yesterday she bent a spoon almost to a U-shape, slightly curved a teaspoon and snapped two sewing needles with the gentlest touch of her fingers . . . Her mother . . . was sceptical at her daughter's excitement, but several weeks and several bent spoons later she has absolute faith in her daughter's mini-bending prowess. The secret is all in concentration, [the girl] says. "I tell it to bend and I think bend. I say it over and over again to myself while I'm rubbing with my fingers. I can't do it always." . . . [The girl] told me she often felt "droopy" after concentrating on bending things. She always felt as if she wanted to lie down . . . Her score to date is three teaspoons, three needles, one dessert spoon, one skewer and two large forks.'

This account contains a number of most interesting features, but let us begin with the problem of how much force her hands were actually applying during the process of stroking. The skewer, the dessert spoon and the large forks take considerable force to bend by direct manual means, even for an adult. It would have been difficult for the girl herself, only eleven years old, to have bent them without using a vice or other mechanical means of applying force. There was no evidence of any marks on any of the bent cutlery consistent with them being bent in such a way, and in any case most of the objects were bent with an adult witness always present.

In these cases only gentle stroking with the fingers was being used. It is still possible to claim that cutlery can be distorted by the use of the fingers alone. This is true to some extent. The amount of pressure that can actually be applied in this way, especially by an adult, *can* be enough to cause a teaspoon or a reasonably thin-stemmed dessert spoon to become twisted. Such a spoon may even bend a little further if it is then left alone for several hours. But it is very difficult to do the same to a thicker spoon or fork, and certainly not to a normal kitchen knife. Far fewer than half the objects which children have bent in this strange way can conceivably have been affected by the force of the child's fingers.

Some objects have been bent which would be well-nigh impossible for the child to bend even mechanically. One is a metal towel rail one quarter-inch in diameter, constructed of mild steel with a chromium plate finish. This had been bent by one twelve-year-old girl through about 40° a certain distance along its length, while her parents looked on. To achieve that

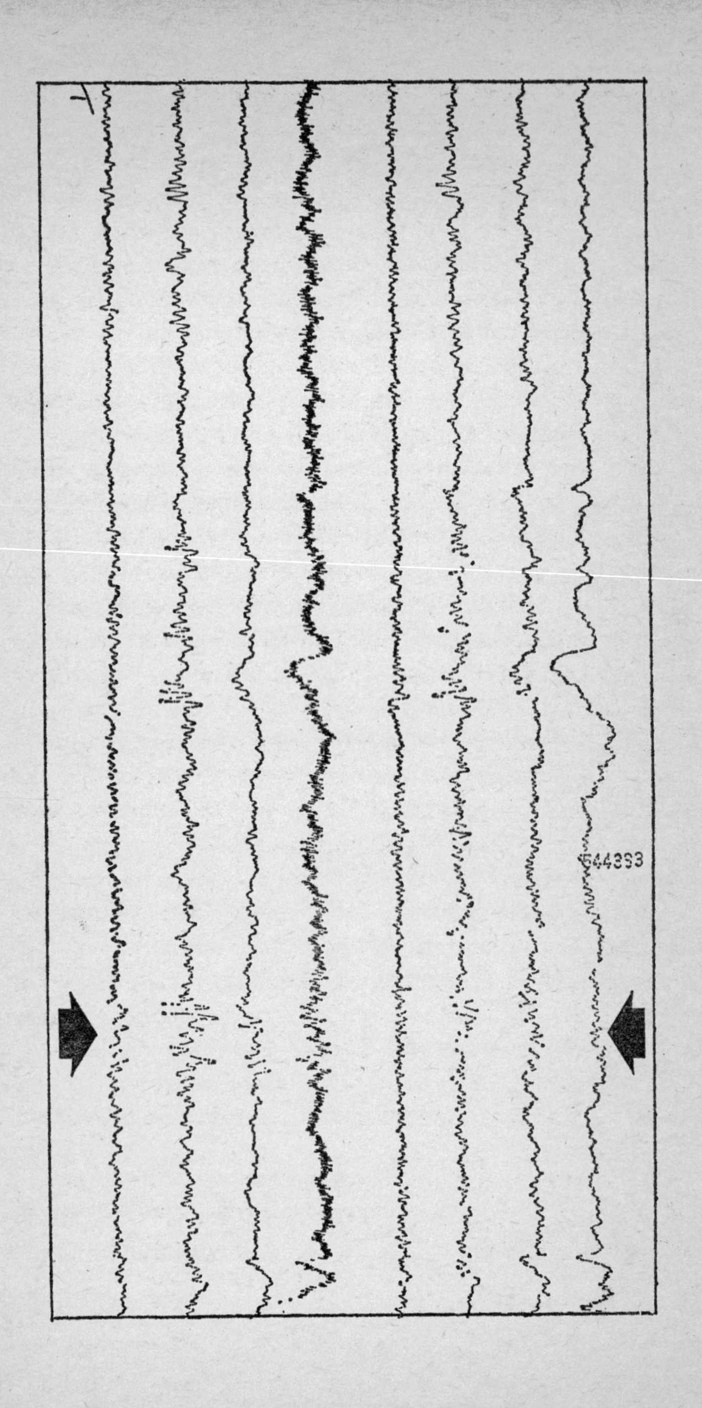

amount of bending would have required a force of at least a quarter of a ton, as was afterwards found by subjecting an identical rail to a measured force. This tallies with the normal yield strength of five tons per square inch quoted in metallurgical tables. In addition there was no buckling whatever of the chromium plate; nor were there any marks on the surface of the rail to indicate mechanical intervention. Again, an eleven-year-old girl bent metal by holding it between the first two toes of one foot and rubbing it backwards and forwards between the corresponding toes of the other foot. One cannot imagine enough mechanical force being exerted like this to achieve the effect!

Other well-authenticated cases have been presented in which metal has been bent without it being touched. This effect was described dramatically by Clifford Davis and Jill Evans at the *Daily Mirror* 'spoonbenders' lunch: 'Fourteen of us sat around the table at London's Hilton Hotel in silence, our eyes closed in deep concentration. After two minutes it happened. A silver-plated coffee spoon, engraved London Hilton 73, curled itself round a saucer. No one touched it. The coffee spoon apparently bent of its own accord.' No wonder the voice of the Hilton's banqueting manager was raised in protest: 'You can have the spoon,' he said, 'but get her [the lady considered responsible] out of here. I don't want all our cutlery twisted.'

An even more bizarre series of events happened to the lady who was thought to be responsible for the Hilton coffee spoon episode. She and her husband and three children all sat round their kitchen table concentrating on bending the metal objects in their hands. They were disconcerted when within an hour various metal objects around them on walls and shelves and in drawers had become distorted. These included a towel rail, ice tongs and two forks. And, even more surprising, the objects had

---

Left: Arrows show the pattern of brain-waves recorded at the moment when metal bends under the influence of a 'supermind', but there is no significant abnormality

all been bent through about the same angle of 80°, and were all at about the same distance from the family group. Was it the mother or one of the children who was responsible?

Then there was the thirteen-year-old girl who laid out some spoons on the floor before Uri Geller was due to be interviewed on a radio programme. When his voice came over the radio she sat and stroked a fork. According to the girl's mother, who sat and watched the proceedings with astonishment, the spoons then bent.

Another journalist tells of a boy he saw bend a key through about 15° by just staring at it. An identical key was tested by a reputable metallurgical firm, who found that a force of 63 pounds was needed to duplicate the distortion. Only a very strong man could have done it. The boy was also observed holding a spoon to his head for five minutes, after which time it started to bend. For both occasions mechanical force can be completely ruled out.

A metal spring is of great value in proving a genuine bending. If a suitably small one is chosen, it can easily be bent by hand, but it always springs back. It is very difficult indeed to give it a permanent bend. But this was done by one subject, and the strength of such evidence for authenticity is undeniable.

The multitude of the accounts makes it clear that we are dealing with a genuine effect which can happen sometimes as a result of direct contact with a subject, and sometimes without it. The main action in the case of direct contact appears to be that of gentle stroking by the fingers of one hand, usually with one end of a piece of cutlery, for example, held in the other. There are cases in which only one hand is involved, one end of the spoon or fork being anchored in the fleshy part of the palm, the stroking being done with the thumb and forefinger. The length of time taken to cause an appreciable bend seems to vary, but it is normally less than thirty minutes and more than two or three; moreover, for a particular child it can vary considerably from one day to the next.

So also the attitude of the child during the stroking process varies considerably. Some concentrate very hard and focus on the object. This may be accompanied by the child mentally repeating 'bend–bend–bend'. There is even one child who silently swears: 'Bend, damn you, bend.' Others, again, may take little direct notice of the piece of cutlery they are gently stroking. As I mentioned earlier, one little girl of seven dreamily sits in front of her favourite television programme whilst the metal object distorts between her fingers.

One curious feature of the bending process is that it appears to go in brief steps; a spoon or fork can bend through many degrees in a fraction of a second. This often happens when the observer's attention has shifted from the object he is trying to bend. Indeed this feature of bending not happening when the object is being watched – 'the shyness effect' – is very common. It seems to be correlated with the presence of sceptics or others who have a poor relationship with the subject.

One thirteen-year-old girl found herself unable to bend a spoon during an interview with reporters from a local paper, yet immediately they had gone she had no trouble in doing so. She also finds the presence of her younger brother greatly reduces her powers. The same reduction occurs in the presence of television cameras. The difficulties all seem to arise from stress produced in the subject, either by the presence of an inquiring eye, be it the eye of the television camera or the human eye, or as a result of emanations from sceptics present. This puts great difficulties in the way of presenting these metal-benders in action to the public; prizes have been offered of as much as £5,000, but no one has apparently yet overcome the sceptical atmosphere created by the adjudicators.

Another feature of the metal-bending process is that it appears to require a large amount of energy: feelings of fatigue usually are experienced at the end of a bending session. We have seen that one eleven-year-old often felt very 'droopy' after concentrating on bending things and wanted to lie down. This same lassitude occurs in nearly all subjects, even to the point of their feeling quite giddy. The eyes of a child can look very glazed after an hour or so of trying to distort cutlery.

The exhaustion, therefore, which follows metal-bending can be regarded as possible evidence for the emission of energy during the bending process. No child has been weighed before and after a bending session, so as yet there are no hard data on this. But it does show that there is an effect here which warrants further investigation. Energy emission by the subject would also be consistent with one of the fundamental principles of physics, energy conservation – the idea that energy can only be gained at the expense of something else. If true metal-bending did occur with little evidence of energy expenditure, then the validity of the law of the conservation of energy would be suspect.

The instances of bending occurring without direct contact assume an ambiguous position. Usually they happen *during* a session of concentrating on bending, but for one twelve-year-old boy this is by no means always the case. Even with a child who may be concentrating while he strokes the metal, the sudden bending can occur when he has just stopped conscious mental effort. It is almost as if the switching-off of attention produces a release of energy which causes the distortion.

A thirteen-year-old girl finds that when she is focusing strongly on one piece of metal, having surrounded herself with various other metal objects and, in particular, safety and ordinary pins, at the end of a twenty- to thirty-minute bending session the objects around her will have become strongly distorted. The shapes into which some of the safety pins have been twisted are indeed very strange and extremely difficult to reproduce by mechanical means.

Taking all this evidence as a whole, only one conclusion seems to be possible: the strange metal-bending phenomenon is genuine. A newly discovered force which is mainly possessed by children is at work; Geller is not alone in his metal-bending powers. And Britain is far from being the only country for metal-benders – they have been discovered in all the countries in which Geller has publicly shown off his powers. With the numbers of subjects now available, it would seem that the required criterion of repeatability had finally been met. Careful investigation of them would mean that the phenomenon could be probed more deeply. Experiments could be designed and carried out almost simultaneously by different groups of scientists. Add to this the fact that new subjects continue to swell the ranks of those we know about and will doubtless carry on doing so, and the possibility of a full scientific analysis of the phenomenon looks very real.

It was clear, at least when I first began my experiments, that they would have to be of a reasonably simple form – especially since all of the apparatus had to be transported to the subjects' homes (and often on the experimenter's back). It certainly would have been premature to design sophisticated equipment before I had some indication of exactly where to look, a more defined starting-point. So the first tests were of a primitive nature.

One of the first steps in this investigation was to standardize the objects being bent. It is all very well observing literally hundreds of bent or broken spoons and forks, but one needs to know their past history – remember Dr Christopher Evans' demonstration with a prepared spoon. Accordingly, pieces of copper and aluminium strip of about 10 centimetres in length, 0.6 centimetre in width and 0.2 centimetre in depth were prepared. Some of these were cut directly from long strips of the material; others were also heated in an electric furnace at 550°C so as to remove, as far as possible, any sources of stress. It also made them softer so that they could either be used for 'warming up' at the beginning of each bending session, or else be that much easier to bend when placed in sealed containers of various kinds.

Another priority in the early stages was to pin down the range of materials which could be bent. Was it just metal, or were other substances susceptible? To find out, variously shaped specimens of plastic, both flexible and brittle, glass, carbon and wood were introduced into the sessions, together with specimens of other metals – lead, tin, zinc, silver, iron, tungsten, as well as the standard aluminium and copper strips. The answer was that all the metals could be bent when in a suitable form, and the best was a strip or cylinder with the same cross-sectional area as our standard strips. Strips of plastic could also be distorted, bending if made of pliable plastic, and breaking if not. On one occasion, a subject succeeded in breaking a carbon rod. Nothing has so far happened with glass.

Then arose the question of the relative ease with which the different objects could be distorted, if indeed it were found possible to do so. The aluminium strips initially proved far more difficult to bend than the copper ones, though one new subject had no problem in bending a very hard aluminium comb completely double. Working with three particular subjects over several months led to a considerable increase in the subjects' powers, so that they could ultimately bend and break strips of aluminium. The copper strips resisted for several sessions, and then were fractured, usually at their mid-points.

Our next step in the investigation was to determine the nature of any radiation which might have been involved during the bending process – with the idea of its possibly being the causative agent. Electromagnetic radiation, described in Chapter 2, is an obvious candidate for this. Whether electromagnetic radiation of a certain range of wavelengths is involved or not can be discerned by making use of the fact that it cannot penetrate metal shielding of a certain thickness. So, if bending of objects could still occur when effectively screened from these wavelengths, then clearly such radiation could be discounted as a factor in the bending process. Ionizing radiation – radiation producing electrically charged particles, or ions – must also be considered as the possible agent in metal-bending. Such radiation can be detected by a Geiger counter, an instrument which we know Geller himself can influence.

In preliminary experiments our subjects were asked to bend strips of softened aluminium, copper or silver which were sealed inside tubes that selectively excluded electromagnetic radiation of certain frequencies. Two of the tubes were of fine wire mesh, effective against electromagnetic radiation of wavelengths in the range of greater than 1 millimetre, and the other was made of quartz which is transparent to visible and long-range waves and not much else. The former type of shielding is already familiar to us from the experiments with Geller described in the previous chapter. The presence of the shield didn't seem to impair his ability to bend a piece of metal sealed inside. Yet this result was not confirmed by my subjects. No one succeeded in bending the strips in any of the tubes during laboratory sessions. It is worth mentioning, however, that two of the best metal-benders, after a week at home with samples of all three types of tube, returned them with the aluminium strips inside fantastically distorted, and none of the seals had been broken. This gave some hint that either very long- or very short-wave radiation was at work, while ordinary radio waves seemed to be excluded.

Direct handling of the strips themselves often had startling results. I saw a strip of silver bend up and flop over on being rubbed gently by one subject. Another subject, who had been left with a strip of copper, proceeded to break it to pieces by producing softening, even going so far as being able to pull off a small piece.

It might be possible that bending might occur as a result of a temperature gradient being set up across the specimen. This would result in differential expansion and therefore in distortion. When railways are built, gaps are left at intervals in the lines to prevent them buckling, so such expansion is hardly negligible. In no case of metal-bending has any temperature increase been noted other than that which could be caused by the warmth of the subject's hand. To test this, a thermoelectric thermometer measured any rise in temperature on the surface of a specimen while bending was actually taking place. This was done in three different instances, and at no time was there any rise in temperature above that which normally comes with heat from the hand. Thus on one occasion a maximum temperature of 31.2°C was recorded at the bend during the bending of an aluminium strip; and the same temperature (to within a degree) was registered when the subject's fingers were resting lightly on the strip.

An electric current flowing through the metal during the bending process could be a source of temperature rise. Even though the tests showed no temperature increase, a current might still be flowing and so giving rise to a difference of electrical potential. So I made a preliminary attempt to detect this with two leads soldered to a specimen strip; one lead was attached to the point of greatest bending and the other to one end of the strip. No potential difference could be detected across the metal strip during the bending, which meant there was no current flow in the strip itself.

The lack of success I had had so far in discovering effects in the metal specimens being bent led me to search for other monitoring devices. The first of these was a heat sensor. The instrument was held close to a specimen as it was stroked, and in six different tests the effect observed when bending actually occurred was less than that caused by movement of the hands across the face of the instrument's detector.

Because of the instrument's shape one cannot position the detector closer than ten centimetres from the bending metal, so that highly focused radiation could still have been present, but not observed. Five excellent subjects then tried in turn to influence the detector by concentrating on it directly and willing it to bend – but not touching it. All that happened was that the axle, a metal part of the instrument, broke!

Geller is supposed to be able to influence a magnetometer with his hands, as we have noted in the previous chapter. So we used such an instrument to see if any of the subjects could do likewise. In three sessions of bending no deflection of the needle was observed – meaning that no change in magnetic field larger than one thousandth of that of the earth could have occurred. Nor were our subjects any more successful in trying, over a period of fifteen minutes, to alter the instrument's reading directly.

Next I decided to test the subjects for ultra-violet radiation, and two simple detectors were constructed. One of these instruments was the same type as that used before on Geller, a strip of softened aluminium covered with material that gives out purple light when radiated with ultra-violet (sodium salicylate) and sealed in a quartz glass tube; several of these were prepared. The other type of detector was a steel plate 10 centimetres wide, 15 centimetres long and 0.1 millimetre thick, covered on one side with a coating of the same material. Five subjects were separately asked to concentrate either on one of the sealed tubes or on the plate while in a darkened room. They were allowed to stroke the tubes gently, but not the plate, since that would have removed the phosphorescent material. None of them produced any effect with the plate, and the amount of visible light in the tubes was no more than could have been caused by the static electricity produced on the glass by rubbing. All the detectors were more strongly affected by a helium tube (which is a source of ultra-violet), held at a distance of 10 centimetres.

A further possible candidate (and I was running out of candidates) was ionizing radiation, as I mentioned earlier. To detect this, I used both a gold-leaf electroscope and a Geiger counter. The electroscope was charged so that the leaf was at a measured deflection, and then a subject tried to bend a piece of metal placed on the other side of the electroscope or a piece held close to the plate of the instrument. If ionizing radiation had been present, there would have been a more rapid leakage of electric charge than normal from the leaf; it was the same as when the electroscope was normally discharging, but then no bending was taking place. Nor could any of the subjects cause the charge on the electroscope to leak away more rapidly if they concentrated on the leaf or plate directly. In other words, no ionizing radiation was in evidence.

The position with a Geiger counter, however, was not so clear cut. Many bending sessions were monitored, either with the Geiger counter which Geller had affected or with a large instrument. On no occasion was there any deflection above the steady background level. The probes were held very close to metal either being bent by stroking, or just being held still in one hand. The absence of any effect here agreed with that observed with the electroscope.

But strangely enough, deflections *were* obtained when the subjects tried to influence the larger monitor directly. Such an effect had already been observed with Geller on the smaller monitor when, as we saw in the previous chapter, he tried to influence it directly. The results with one sixteen-year-old girl, whose metal-bending powers were not well attested, were startling. According to the record on the strip chart, she achieved a series of huge changes in the reading on the monitor: the biggest being up to about *a thousand times* the background level. A very curious situation had arisen. The instrument was in good working condition, yet could be affected by human beings in a way which almost suggested a malfunction, as we could be nearly certain that no high energy radiation was in fact involved.

To summarize this preliminary series of tests, there was no concrete evidence for any involvement of electromagnetic or ionizing radiation during the bending process. Nor was there any indication of temperature rise or electrical current flow in the specimen being distorted. But what could now be said with complete confidence was that the Geller effect does occur, and can also occur at a distance. Apart from this, the one positive finding was that metal-benders can make Geiger counters malfunction. It was evident that a number of the pieces of the jigsaw puzzle had been assembled, but it was not clear that they belonged to the same puzzle. The time had come to try to make sense of them in the light of the body of scientific knowledge.

5 Explanations

The Geller effect cannot be lightly dismissed. People across the world can cause a range of objects, especially metallic ones, to bend into strange shapes and even to break, using a method which is apparently beyond our current scientific understanding. Yet we are compelled to try to give a scientific explanation. We may not ultimately succeed in this aim, but only when all avenues have led us nowhere will science finally have been proved wanting.

Before we consider what mechanisms could possibly underlie the Geller effect, we must first help ourselves by starting with a brief summary of what it is that we shall have to try to explain.

Let us take the crucial feature: the bending itself. It is often found that metal objects may become considerably distorted – sometimes acquiring very weird shapes. The handle of a spoon may twist round like a corkscrew, or the whole spoon may loop around itself in a complete circle. The prongs of a fork may splay out, each in a different direction. A strip of copper may twist round so sharply at the point where it is being gently stroked as, in one case, to give a sharp pinch to the subject's finger. (Oddly enough, once the finger was released, additional bending took place in the opposite direction.) A whole collection of metallic objects, all at about the same distance from a subject, may be bent through the same angle. At other times the amount of distortion caused may be so slight as to be only just noticeable.

We must realize at the outset that there are three different processes involved in the Geller effect, being the three links in the chain from subject to object. The links are, respectively, the generation, the transmission, and the reception of the energy which effects the distortion. Accordingly, there are at least three separate mechanisms to be discovered. There may be more than a single way in which one of the links is being activated, so that several alternative possibilities will have to be considered.

The most important physical feature involved in bending is a transference of energy, from the subject to the object. For only thus can work be done on the object to modify its shape, and work requires energy. The energy transfer could occur in a number of ways. Where bending occurs by stroking, energy could be directly transferred from the fingers of the subject to the object, and the direction in which distortion occurs could be determined by the pressure applied by the fingers. However, the degree of warping actually observed often seems to be far greater than could arise by simple muscular action. Also, the behaviour of the copper strip which bent first one way, pinching the subject's finger, and then the other way, though the stroking movements were always made in the same direction, shows that the pressure of the fingers alone cannot determine the direction of distortion when it occurs by stroking.

It is obvious that when the Geller effect occurs at a distance from the subject, the energy must be transferred by other than direct contact. It seems that the transmission process can produce focusing of energy on a quite small area, say a few square centimetres, since objects a few centimetres away from a bent object may not themselves become distorted. Only if the energy is translated in unequal amounts to different portions of the specimen might we expect bending – the distortion resulting from various parts of the object being forced to move in different directions. An unequal distribution of the transmitted energy must, as in the case of the copper strip discussed above, sometimes have arisen when bending occurred by direct contact.

I have myself directly observed the Geller effect in objects made of a range of metals: silver, lead, zinc, tin, copper, aluminium, iron (including various types of steel) and tungsten, with bending of up to 180°. Objects made of iron, copper and aluminium have been broken as well. The specimens tested had lengths varying from 5 centimetres to 30 centimetres, and thicknesses in cross-section varying from 0.1 millimetre for thin strips of copper or silver to rods of mild steel of 0.6 centimetre in diameter. When working with specimens of iron, copper and aluminium of comparable size, subjects seemed to find the aluminium ones hardest to bend to start with. Thicker objects were harder to bend than thinner ones, but the relationship between the length of the specimen and the ease experienced in bending it was not found to be a simple one; and we will return to it later.

Plastic about 10 centimetres long and in cross-sectional area 0.1 to 0.5 square centimetre thick can also be caused to bend or break. Specimens of flexible plastic have been bent without breaking and brittle plastic strips have been fragmented. Once when a strip of brittle plastic was being rubbed gently it exploded into five pieces, one piece actually hitting the ceiling. There is also one instance of a specimen made of soft plastic being bent without contact.

Various other materials have also been investigated. My subjects have found difficulty in Gellerizing objects made of wood, glass or china, though there are reports of other subjects having shattered objects made of wood and glass. Geller himself claims he can cause glass to bend, but this has not been substantiated by any specimens. Several subjects caused small carbon rods to break but no bending was noticed before the fracture occurred. Some, like Geller, have also cleaved single crystals.

It is possible to estimate how much energy would theoretically be needed to bring about the distortion in objects which have been Gellerized. Let us consider, for example, a strip of aluminium 10 centimetres in length, 0.5 centimetre in width and 0.2 centimetre thick. The total energy necessary to distort it into a semi-circle can be calculated to be about ten watts if the bending occurred within one second. This estimate agrees very well with results obtained from metallurgical tests. Obviously more energy is needed to bend thicker strips than thinner ones as there is more metal to be bent, or if a strip is shorter rather than longer, because the degree of curvature in bending is then increased. In the metallurgical tests, this amount of energy was usually

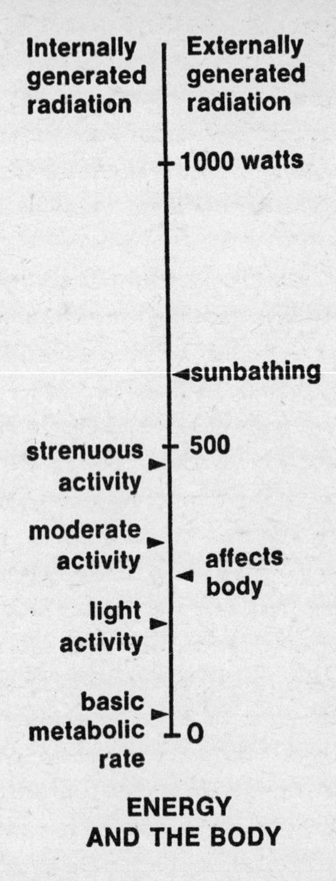

ENERGY AND THE BODY

delivered over a period of about five minutes, so that the power being transmitted to the material must have averaged about 30 milliwatts (a milliwatt is a thousandth of a watt): this amount of power seems tiny if we compare it with the 60 watts consumed by the average electric light bulb.

Man has had to set up safety regulations to protect himself and his unwitting fellows against the ever-widening range of radiation produced by the gadgets of modern technology – the hazard of X-rays has long worried us, and recently it has been found that thousands of unsuspecting Americans have been absorbing them at a dangerous level of intensity from their colour television sets. Microwave ovens have been shown to emit microwave radiation (another form of electromagnetic radiation) and there is the quite horrifying story, perhaps apocryphal, of the man who was killed by the Distant Early Warning Radar System, which also uses microwave radiation. There was not a scratch on his body, but an autopsy showed that his internal organs had been literally fried.

Because of the danger, acceptable safe levels of radiation exposure have had to be determined very carefully. In the West the safe level specified for microwaves is 10 milliwatts per square centimetre, though in Eastern Europe it is one thousand times less than that. In the metallurgical experiments the amount of 30 milliwatts required to bend our aluminium strip had to be distributed over an area of 6 square centimetres, and this needed a level of power directed on it of 5 milliwatts per square centimetre – a level dangerously close to the maximum permissible exposure level just cited. To bend a strip twice as thick would require power at five times the safe level. Only if the subject could focus the power emitted on to the specimen being bent could there be any reduction in total power emission. If focusing does not occur, the total power being radiated by a subject who had been able to bend our strip at a distance of a metre would need to be at least a kilowatt, if not considerably more. If this were microwave power then the person would be like an electric fire. In any case, the total amount of power consumed by the activities of brain cells has been calculated to be somewhat less than a hundred watts, so it is hard to see how the brain could produce an output which vastly exceeded this.

Another feature of the Geller effect is the bending of metal inside certain types of container. If the effect were transmitted by electromagnetic radiation with a wavelength greater than 1 millimetre it could not have penetrated the wire mesh shields used in the various tests with much less than a tenfold reduction of intensity. They would not, however, exclude electromagnetic radiation of considerably longer wavelength, so that the long wave (or low frequency) end of the electromagnetic spectrum remains a possibility along with electromagnetic waves with a wavelength of less than a millimetre. Thus most parts of the frequency spectrum of electromagnetic radiation – low frequency, microwave, infra-red, visible, ultra-violet, X-rays and gamma-rays – could all be carriers of the bending power.

A distinctive mark of the Geller effect is that it has what might be called an 'intentional' aspect. Specimens only become distorted when a mental effort is made by a person that they should do so. There have been a few reports of objects becoming bent when no nearby subject was concentrating on them, but they are not at all well attested, and the effects may have happened through previous contact with a subject. All authentic cases of the Geller effect would appear to require in the subject a suitable frame of mind. The amount of conscious attention may not be very high, but for the most powerful of the children I have worked with, and certainly for Geller, intense mental effort accompanies the effect.

Finally, it is apparent that the Geller effect extends to a wider range of psychokinetic phenomena, involving electronic equipment. There also appears to be control of physical processes at the atomic level, in particular that of radioactive decay, as seen, for example, in Geller's influence on the Schmidt machine described earlier.

Let us touch for a moment on the subject of telepathy. Communication of shapes and sounds from one person to another does seem to occur, even, so it appears, over distances of thousands of miles. The rate at which information is transmitted does not seem to be very high, but quite complicated patterns can be sent. Signalling by these means can occur through walls of houses as well as sheets of metal. Furthermore, some subjects with metal-

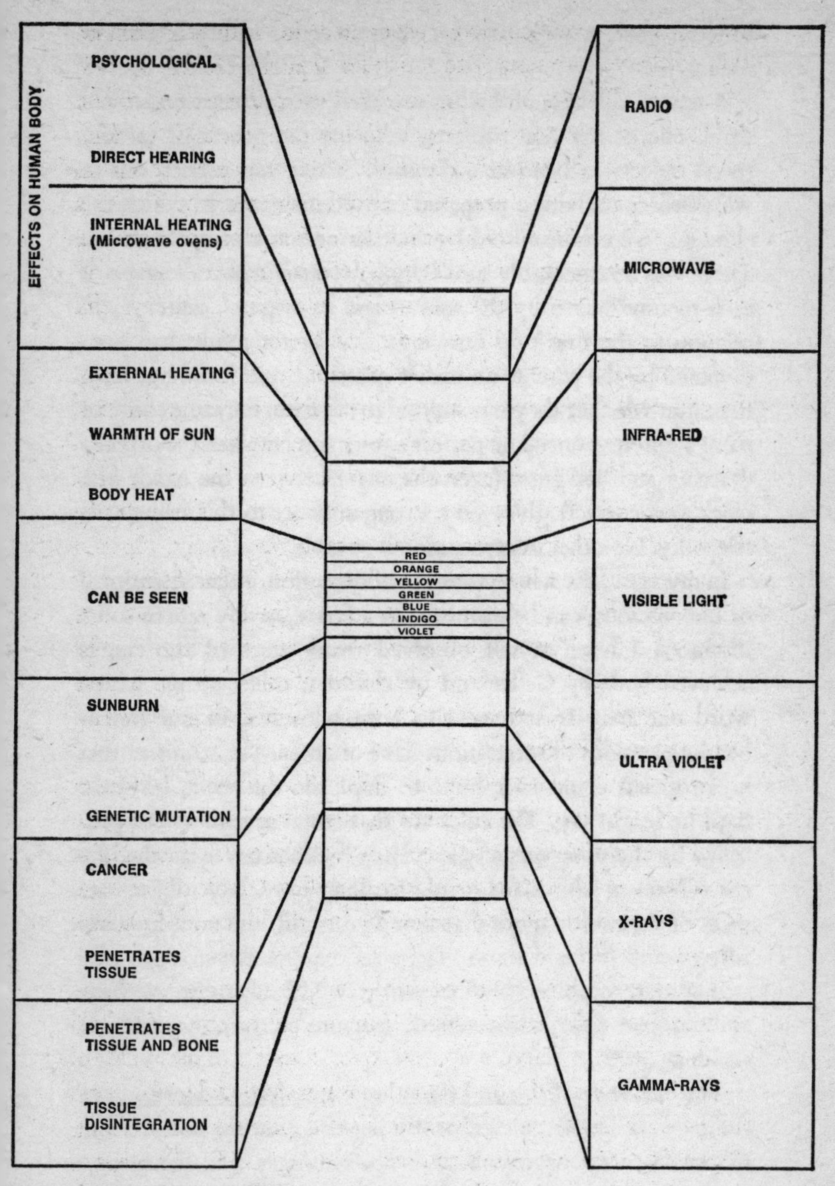

The electromagnetic spectrum

bending powers appear to have a heightened ability to receive such signals.

A remarkable case of this has occurred with a thirteen-year-old child, one of my best subjects, who has the power of causing metal objects to bend at a distance. A test was carried out in which three drawings, prepared outside the room in which the child sat, were transmitted by another person also in the room. There was a remarkable similarity between the overall shape of each picture drawn by the subject and its original, especially in relation to the first two drawings. Still better results had been obtained by the parents on earlier occasions, and it did not seem to matter whether they transmitted to her from the same room or from a different one. The perfect agreement between the original drawing and her guess (even the angle between the hands of a clock was correct) gives very strong support to this being true telepathy. No other interpretation is possible.

In my view, the whole question of deception, either intentional or unconscious, can be dismissed as a factor, at any rate in some instances. I have myself observed many cases of the results achieved both by Geller and by children; other people whose word one may trust have also been witnesses. In our metal-bending sessions the conditions have often been so stringent that no magician could ever hope to duplicate the result, however hard he might try. The rules are that metal specimens are provided by the observer; any specimen is either never touched by the subject or else is for gentle stroking only. My subjects are most of them little more than seven years old, and none are over fifteen; and to bend some specimens requires the strength of a strong man combined with a small vice. What is more, we have at least one fully authenticated instance of bending without touching at all.

The existence of this and the other examples of distortion at a distance implies some mechanism for transmitting power from the brain to the specimen. It is not necessary that the energy required to produce the ensuing warping come from the brain

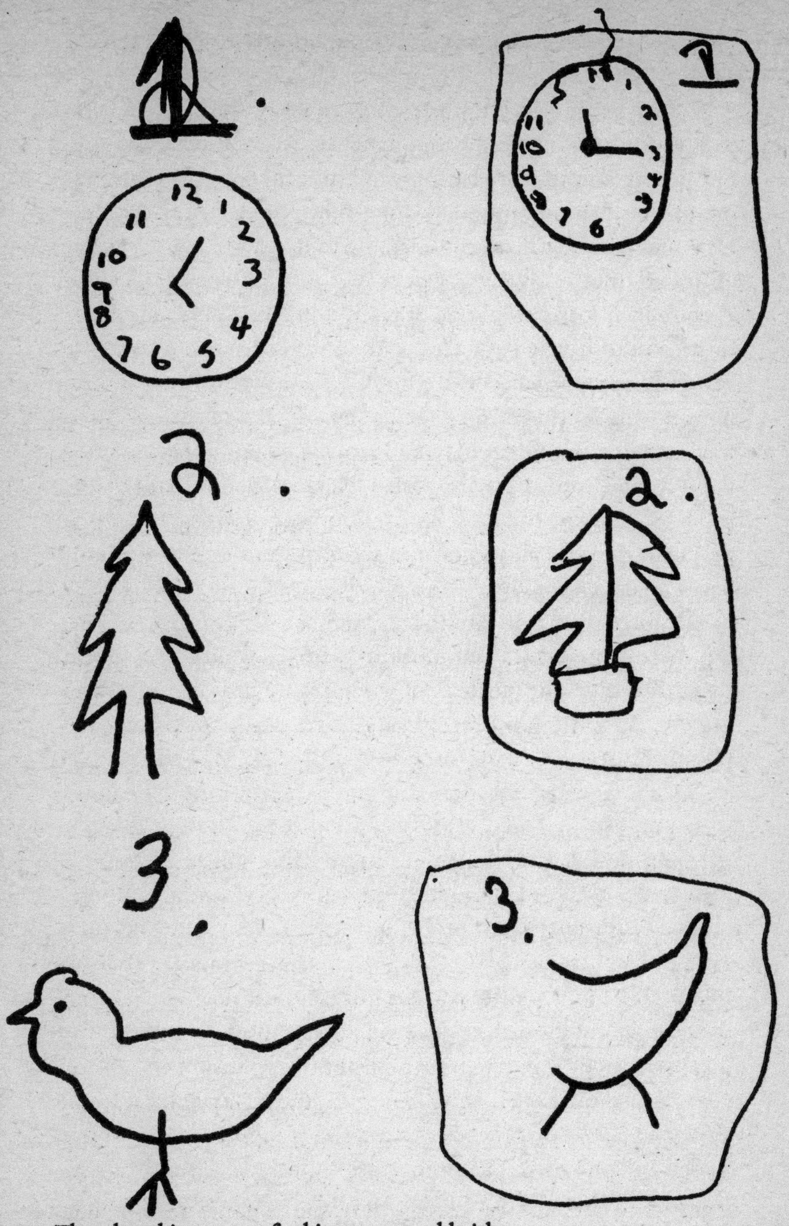

The telepathic success of a thirteen-year-old girl

itself, but the intentionality of the subject – he or she needs to be 'willing' the bending to occur at some stage – must be propagated by some means to the vicinity of the object.

We now have a preliminary picture of what the Geller effect is, but the account that has been given is not enough to let us single out the mechanism responsible for achieving the effect. Yet there are enough data at hand considerably to narrow down the range of possibilities. However, before we turn to consider in any detail which forces might be of interest in metal-bending, it will be helpful to have a clear idea of the forces of nature as they are generally recognized. For it is among these forces that we must hunt for the mechanisms of metal-bending and telepathy. More than one of these forces may be involved. It is also possible that the questions posed by the Geller phenomenon will never be resolved in terms of any force we know of already; but at the moment, they are all we have to work with in terms of present scientific understanding.

The forces of nature are four in number – electromagnetism; the force between the sub-nuclear particles which holds them together inside the nucleus of each atom; radioactivity; and gravity. As we might expect from the enormous complexity of the world around us these forces have a large range of effects.

Electromagnetic radiation comprises energy propagated at a speed of 186,000 miles (300,000 kilometres) per second in a vacuum, and has, as stated earlier, a wide range of possible wavelengths. Thus long radio waves of, say, ten thousand kilometres' wavelength can occur, as well as radio and television waves with wavelengths of between one centimetre and ten kilometres. In the shorter wavelength range are microwaves with wavelengths of the order of one millimetre; infra-red rays of, say, a thousandth of a centimetre wavelength; visible light of roughly a tenth of the wavelength of infra-red (though lying in a much narrower band of frequencies); ultra-violet rays ranging from the violet end of visible light to rays with a wavelength a ten-thousandth that of visible light; and finally X-rays and gamma-rays, which have as small a wavelength as one chooses below that of ultra-violet light. The basic effects of the different types of radiation can be seen on page 89.

The force of electromagnetism is a combination of the forces of electricity and magnetism. This synthesis arises because electrically charged particles at rest repel or attract each other, depending on whether their charges (negative or positive) have the same sign or the opposite one. If they are in motion they generate further magnetic fields which can modify their interaction. Thus magnetism can be regarded as electricity in motion. The chemical properties of matter are determined by the number of negatively charged electrons which can circulate in an atom around a heavy central nucleus with a positive charge. For it is the process of sharing or swopping electrons which binds atoms into molecules and holds molecules together. Thus the properties of the various forms of matter – crystalline, metallic, glass-like and so on – are

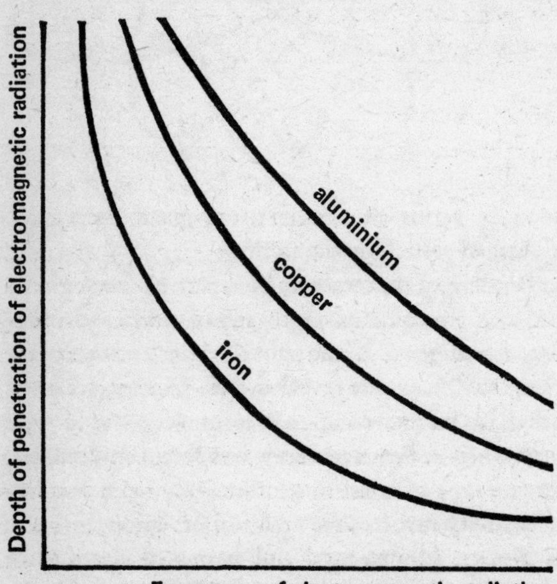

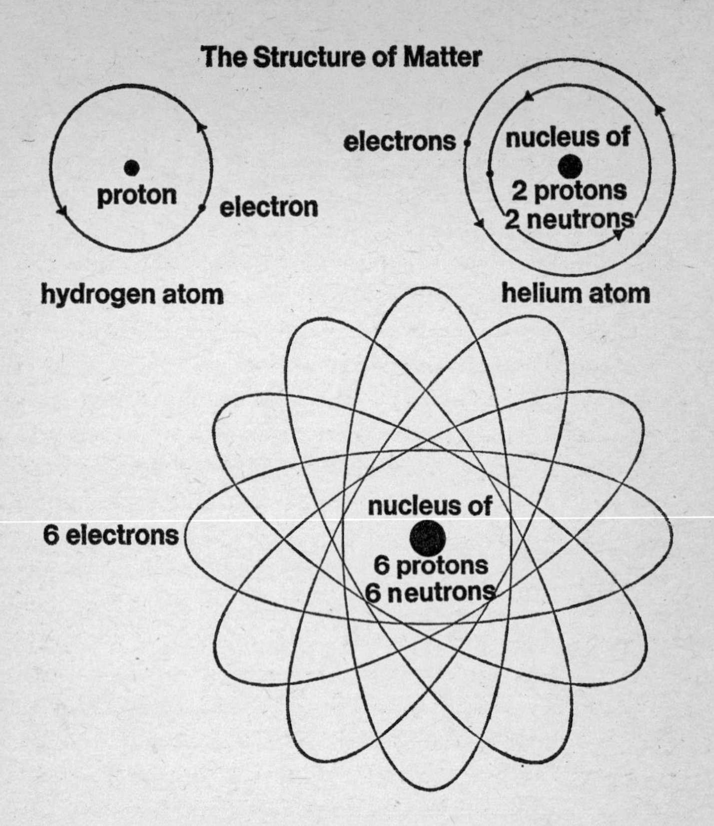

all understood in terms of the electromagnetic interaction between the charged particles in the atom.

This understanding of the structure of matter has been gradually expanded as more complex aggregates of atoms and molecules have been investigated. At the same time the structure of the atomic nucleus itself has been revealed with increasing clarity. That the nucleus has internal components was realized at the turn of the century when radioactive decay was found to involve a change from one type of atom to another – say from uranium to lead. Radioactivity involves the emission of various forms of radiations – X-rays, gamma-rays, and beams of electrons or other particles. There may even be emitted fragments of heavy nuclei which have disintegrated into lighter ones. And clearly the fission of a nucleus can only occur if it has component parts.

From radioactive decay it has been discovered that every atomic nucleus is made up of two distinct building blocks. One is a particle about two thousand times heavier than the electron and with an equal but opposite charge, called the proton. The other has about the same mass but is electrically neutral, and is termed the neutron. Each nucleus contains whole numbers of neutrons and protons confined to a volume which is about a thousand-billionth that of the atom as a whole. To achieve such confinement there must be a very strong force between these subnuclear particles, protons and neutrons. This nuclear force must also be strong enough to overcome the electrical repulsion between the protons held so close together inside the nucleus. Both this force and the electromagnetic one are much stronger than the force producing radioactive decay, which can split particles up millions of times slower than the former two forces.

There is finally the force of gravity, which is one of attraction between all particles. The strength of the force is proportional to the mass of the interacting particles. The attraction only becomes appreciable when very massive objects are involved such as planets or stars. Thus gravity is unimportant in determining the interaction between two man-sized objects; at least one of them must be a heavenly body.

These four forces of nature, in spite of their differences, all affect us here on earth. As well as producing the tides through the pull of the moon, the force of gravity holds us and all things on the earth. It has long been recognized as crucial. But now the importance of radioactivity has also been realized as crucial, allowing us to tap the power of the atom in controlled nuclear fission. It also affects us by producing ionizing radiation of various kinds, the by-products of natural radioactive decay. Ionizing radiation has been disturbing and altering the hereditary makeup of living cells over past aeons, causing the mutations as a

	Strength	Range	Effect in Humans
Electromagnetic	$1/137$	∞	Many, from behavioural changes to damage to nervous tissue
Gravitational	10^{-40}	∞	Attraction to the earth or moon. Absence causes muscular degeneration
Nuclear	15	10^{-13} cm	None directly. Beams of nuclei cause radiation damage, leading to death
Radioactive	10^{-5}	10^{-14} cm	Causes radiation damage and death

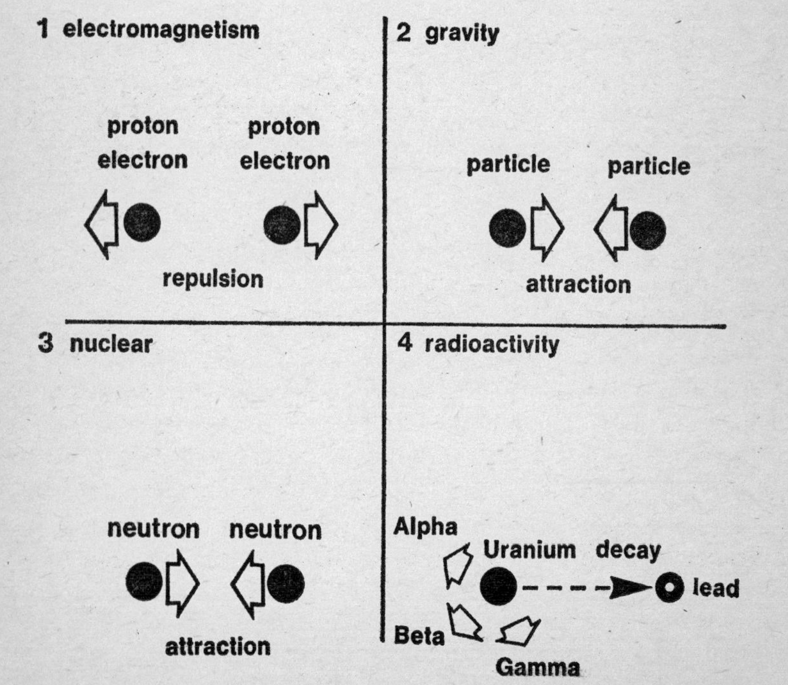

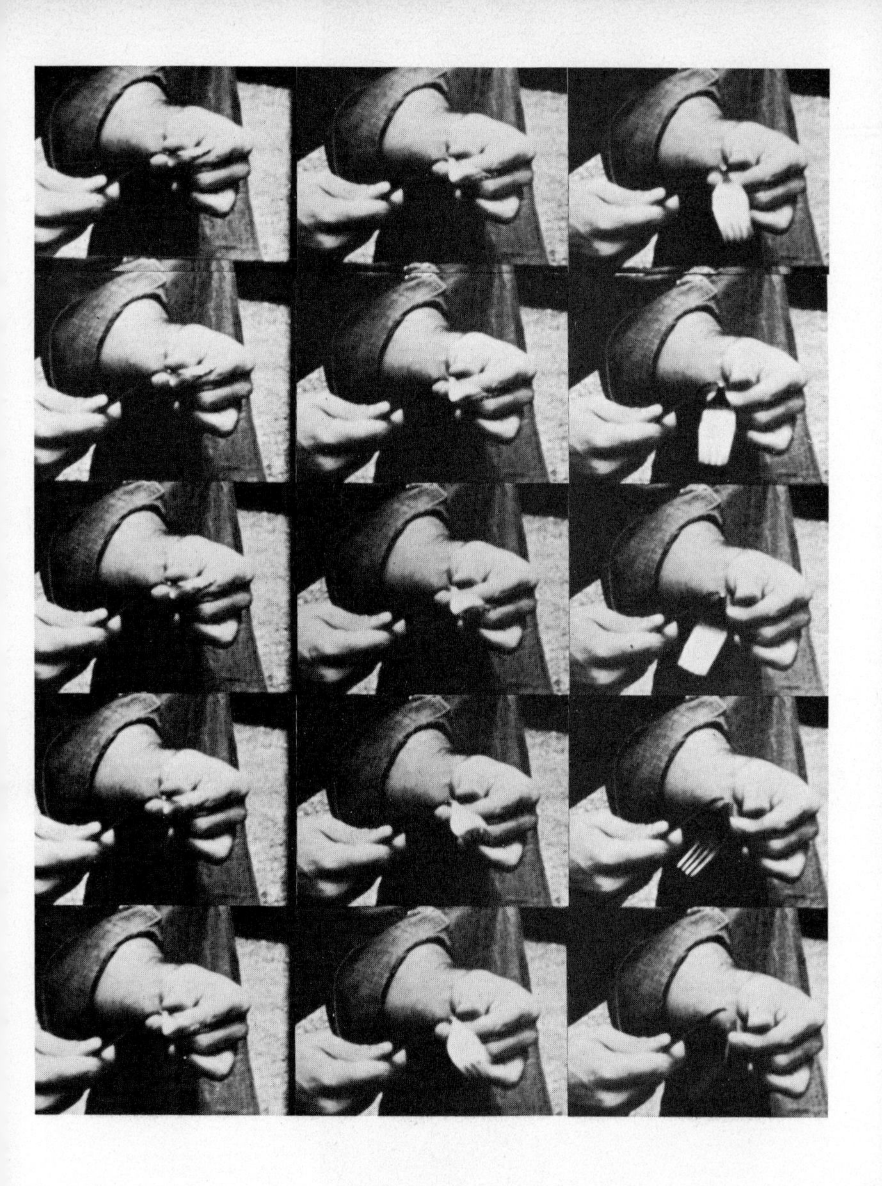

A remarkable sequence from a film made by James Bolen of *Psychic* magazine, showing Uri Geller bending a fork. As Geller stroked it between his thumb and index finger, the fork gradually became pliable at its mid-section. (*Psychic* magazine)

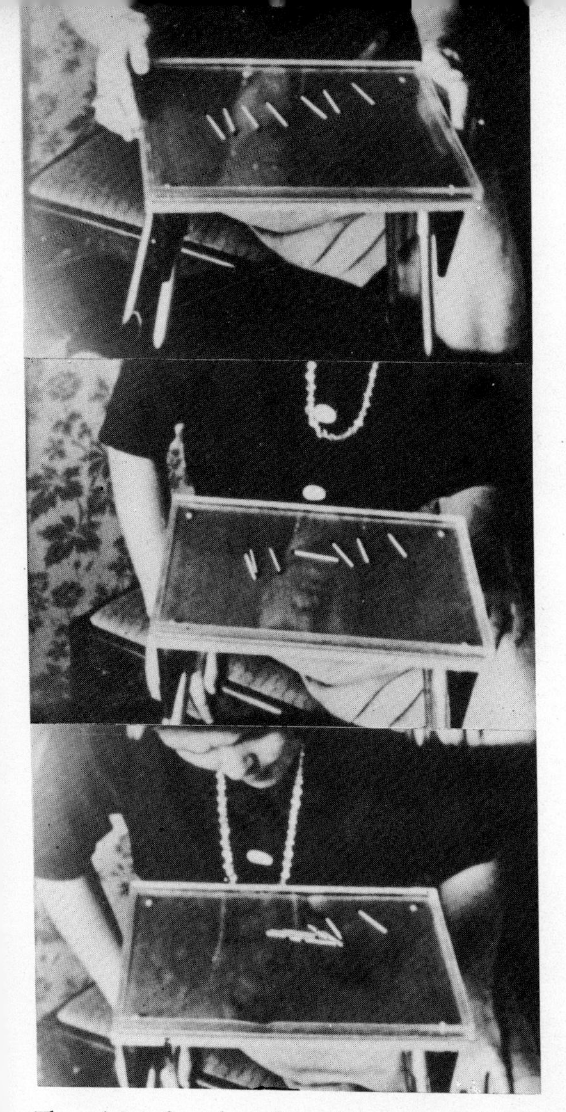

The activity of matches influenced by the psychokinetic powers of Madame Vinogradova. Her hands had no contact with the table after the initial picture, according to scientific observers. (The Benson Herbert Paraphysical Laboratory)

left A paperknife comes to the end of its useful life under the influence of Uri Geller. (Topix)

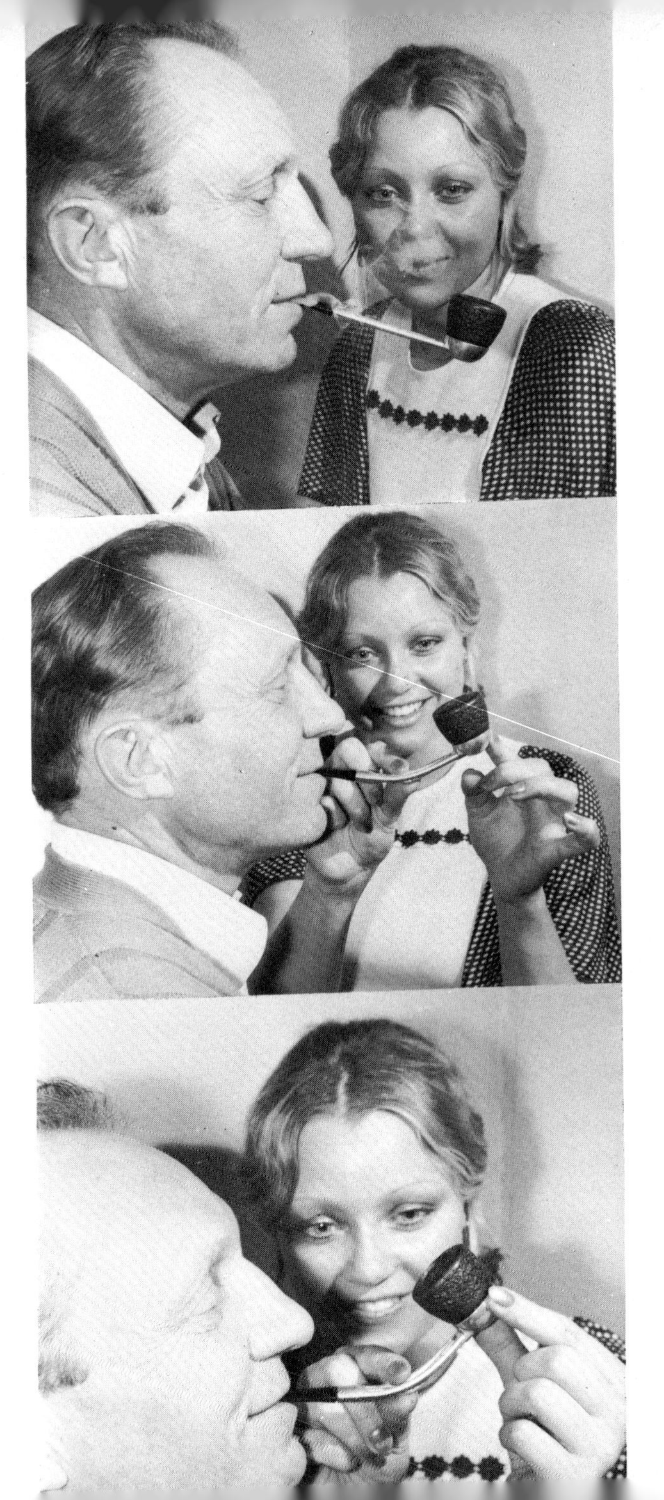

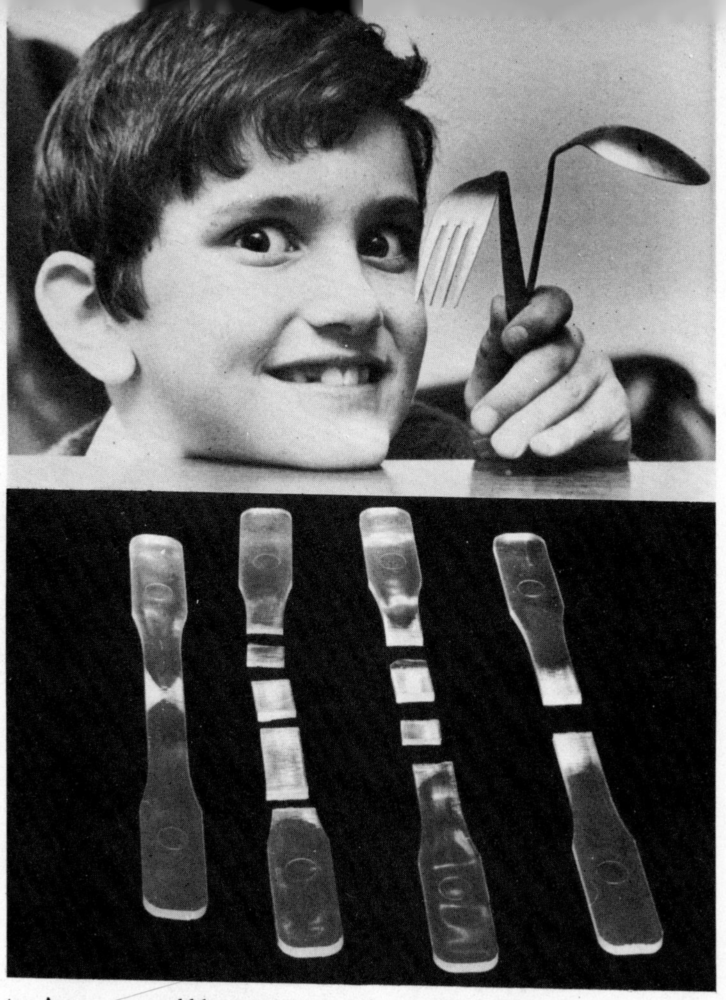

top A seven-year-old boy with his handiwork. (Syndication International)
above Clear plastic fractured at various points. The white lines show where stress is greatest. (Macmillan London)

left A young lady concentrating on bending a pipe *in situ*.
(Syndication International)

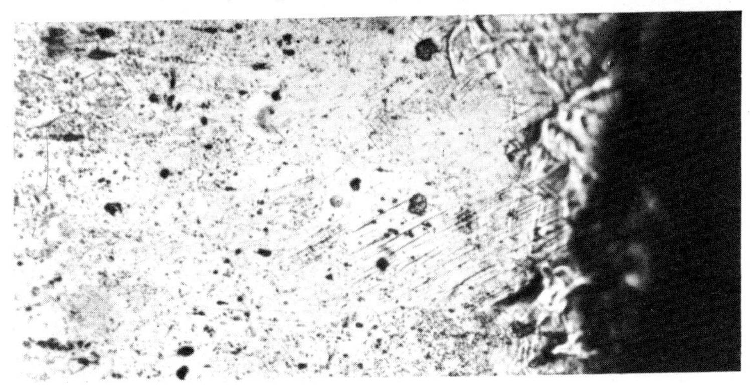

Details of the deformed grains near the fractured region in a mechanically torn strip of copper. The striations are regions where slip has occurred.

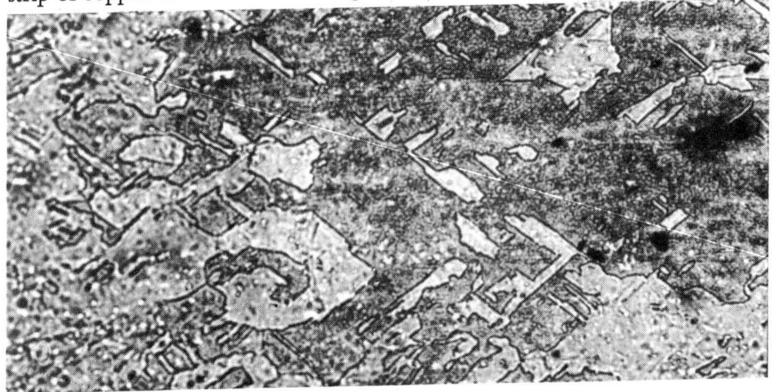

Appearance of the crystal grains in an undeformed region of a copper strip. The edges of the grains are shown clearly by the direct lines and they are undistorted.

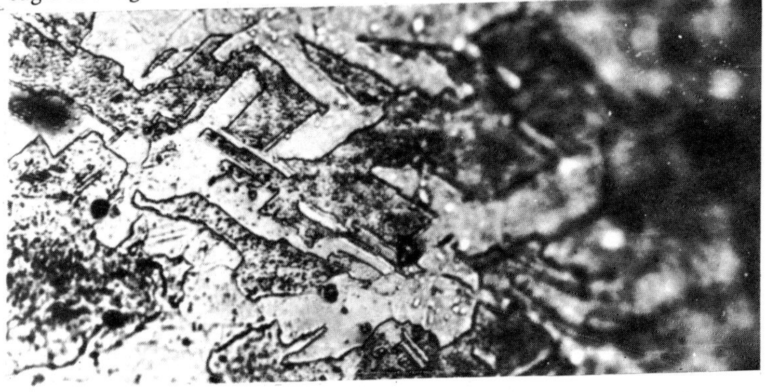

The surface close to a fracture caused by gentle stroking of a copper strip by a subject. The crystal grains are undeformed right up to the main fractured surface in the middle of the picture.
Magnification x 800. (Professor John Taylor, courtesy BBC)

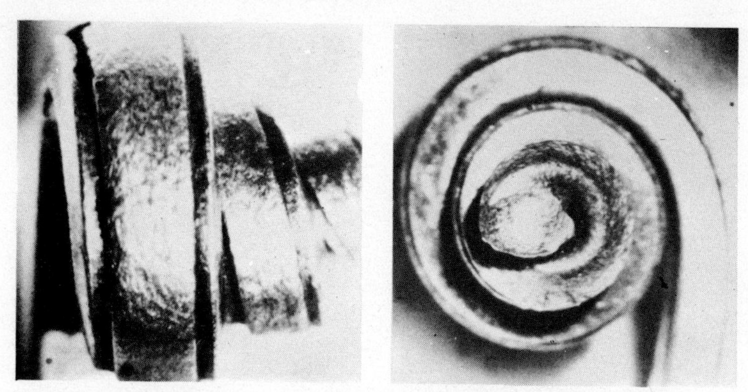

Two views of the handle of a silver teaspoon coiled by psychic powers. (Macmillan London)

Metal sculptures created by a ten-year-old boy's psychokinetic powers. (Professor John Taylor)

Magnification x 10,000 of a spoon handle fractured by stroking, showing possible indications of local melting. (Professor John Taylor, courtesy BBC)

result of which we, for example, have evolved from primitive single-celled creatures into human beings.

Electromagnetism, being the basis of all chemical properties, clearly exerts an all-pervasive influence on our total environment. But it also has a very direct influence on living matter, as, for example, the effects of radiant heat (a form of electromagnetic radiation) show. Other varieties of electromagnetic waves, such as X- and gamma-rays, are ionizing, and so can also cause mutations, as well as producing other damage in living cells. Microwaves in turn cause heating, while radio waves are thought to produce psychological effects by acting on the brain.

The nuclear forces are also of great importance to living matter. They ensure stability of the nuclei of which all living things are composed, as well as generating life-giving energy through nuclear fusion taking place in the sun. At the same time, distant stars which have used up all their available nuclear fuel sometimes find themselves a little too heavy to settle down without a strict diet to a graceful old age as white dwarf stars – about as big as the earth. This loss of weight has to be accomplished rapidly or the star will collapse completely in on itself owing to the gravitational pull between its parts, and so produce a black hole. The only way to avoid such a fate – clearly one worse than death – is for the star to eject the outer part of itself. When this happens, violent explosions are thought to be visible as supernovae, stars which suddenly flare up becoming for a few weeks as bright as their own galaxy. In the process, some of this ejected matter is violently accelerated and is thought to contribute to the cosmic rays – fast-moving protons – which are continually showering the earth. These also can affect living cells and cause mutations. It has been conjectured that periods of great evolutionary change could be related to nearby supernovae explosions causing an increase in the mutation rate.

Left: The four forces of nature and their properties. The strengths in the top table are pure numbers

Of all the forces which we have been reviewing in a very general way, it is electromagnetism which has contributed most to technological advance. We know more about it than we do about the other three, and it is with this force that we shall be mainly concerned in trying to establish criteria for dealing with the Geller phenomenon.

Electromagnetism is the most attractive of the various candidates for the 'intentionality field' surrounding a person trying to bend metal. The range of wavelengths of electromagnetic rays is enormous. There is hope of fitting the transmission of 'intentionality' somewhere into their span. What makes this possibility even more likely is the rapidly increasing understanding of electromagnetism and its interaction with matter. There is certainly a growing appreciation of the role electromagnetism plays in living tissue – the curative effects of negatively charged particles in the air and of low frequency currents on diseased organs, the use of electric current as a diagnostic tool, the effects of the earth's magnetic field on the growth and responses of organisms, and electric shock treatment for behaviour diseases. Electromagnetism may contain one of the mechanisms at work in the Geller effect. It must, in any case, be involved at certain stages of the process, since the intentionality field can only arise in some fashion from brain activity; this latter involves electrical action in the passage of electrical activity along the nerve fibres in the brain. At the other end of the chain the bonds holding the molecules together inside a specimen of bending metal are electromagnetic in origin. If the object is to be distorted its electrical cohesive forces must be disturbed in some way.

At this point let us recall the words of the Russian scientist Vasiliev. 'We were all certain it could be nothing but radio waves,' he wrote after his telepathy experiments. 'We could not imagine any other explanation. For we all knew there are electric currents pulsating in the brain; and every alternating current sets up its electromagnetic field or radio waves.' But as we saw back in Chapter 2, some of his experiments made an explanation of telepathy in terms of electromagnetic force seem very unlikely. And there is the very difficult question of how the very low energies emitted by the brain as electromagnetic radiation could have any effect on another person's brain. This question is even more critical where metal-bending is concerned, since more energy would be expected to be needed to bend strong iron bars. These problems have caused the electromagnetic hypothesis to be looked upon with disfavour by those interested in such problems. We will fall in temporarily with this conclusion, and for the time being attempt to develop some alternative explanation.

One source of energy which could be considered is that of cosmic radiation. As I described, this consists primarily of high energy protons entering the earth's atmosphere from outer space. If these particles could be focused, then a reasonable amount of energy might be supplied to a piece of metal or other object. This is in close analogy to focusing the rays of the sun by means of a magnifying glass so as to cause a paper or thin wooden object to char and even to burst into flames. It is difficult to conceive exactly how much focusing could occur with cosmic rays, but some form of magnetic field generated by the subject might be possible.

There are radiations in nature other than those of electromagnetism and of cosmic rays, such as beams of neutrons, of neutrinos (massless neutral particles usually released in radioactive decay) and of the various particles associated with the proton, neutron and electron, which between them constitute the atom. Artificially created rays of such particles are now in frequent use to probe the internal structure of the elementary particles, and have been conjectured to occur under certain conditions in the cosmos. Here we find objects constructed of matter in an unexpected state. Stars composed almost entirely of neutrons have been discovered in the past seven years; there is evidence of even heavier stars which disappear to make the gaps in space which we call black holes. Our ideas about matter are being upset. Some fundamental biological concepts, even, are changing so rapidly that what would previously have been regarded as way-out, esoteric ideas are now being seriously considered. It is difficult to see how any of these new potentialities could be used for the brain power required to achieve the Geller effect, but since rather little is known about cerebral function they cannot be dismissed out of hand.

We must turn finally to the distortion brought about in the metal specimen itself. There are various conjectures as to how this might arise, apart from through direct muscular action. One idea is the production of a temperature gradient across the metal, so that one side of it is hotter than the other. The hotter side will

particle	*symbol*	*where found*
electron	e	orbitary nucleus in atom
proton	p	in nucleus
neutron	n	in nucleus
neutrino	υ	emitted in radioactive decay

therefore expand further and result in a bending effect. Another candidate for speculation is some kind of chemical which can penetrate the metallic surface, producing cracks which weaken the specimen and finally cause it to fracture. This is not completely impossible, and indeed perspiration from the hands can be very destructive on brass, for example. Some players of brass musical instruments have to wear gloves to protect the keys of their instruments from corrosion.

A further possibility is that vibrations may be set up in the specimen, either by direct contact or, when distortion occurs without contact, by the intentionality field. It is known that vibrational effects on materials can be very strong. For example, a glass tube one inch in diameter can be pulled apart by means of vibration set up along it. This can be done by making a sharp scratch round the glass and then pulling the moistened fingers over the glass to set up vibrations. Such oscillations can even produce heating over tiny areas when material is worked to such an extent that, for example, ebonite, which is a relatively soft material, can cause the edge of a steel tool to melt. The ebonite itself need not become hot, the temperature rise occurring over a distance of less than a hundredth of a millimetre from the edge of the tool.

Finally, we cannot at this stage exclude forces at work of which we have at present no knowledge. At the end of the last century a famous scientist advised young men not to go into scientific research since all had been discovered; according to him all that was left was to add further decimal points to the existing knowledge of nature. We now know how wildly wrong he was. The atom has been split, the nucleus discovered, and that in its turn cracked open. Einstein has shown how space and time labels can be distorted by strong gravitational forces. The other three forces of nature have been deeply probed, but many puzzles about them still remain unsolved. The paradox currently presented by black holes, and especially by the disappearance of matter at the centre of these 'heavenly' objects, highlights our similar lack of understanding of gravity. With such uncertainty abounding further forces may well be discovered; the Geller effect may be the first indication of their existence.

It is already possible to draw up a short-list of candidates for the Geller effect. It would be best to leave the question of how the intentionality field is generated by the brain until we have got a clearer indication of what the field could be in physical terms. Let us turn, therefore, to the middle link of the chain, that of the mode of transmission of the energy from subject to object.

The other forces of nature besides electromagnetism – the 'strong' force holding the sub-nuclear particles together, the 'weak' force of radioactivity, and the attractive force of gravity – still require careful consideration, but we will not go into this here. We can, however, quite rapidly cross off the use of cosmic rays as being impractical. The amount of energy carried by cosmic rays impinging on the earth is about equal to that carried by starlight, and is therefore clearly very weak. To provide the energy to bend the strip of aluminium described at the beginning of this chapter and needing about 6 milliwatts per square centimetre, a ten-millionfold concentration of the energy of cosmic rays or starlight would be necessary. This means that all the cosmic radiation falling on a square with a side of one hundred metres would have to be collected and focused on to the strip. To achieve this for starlight without the intermediary of a very large optical lens would be impossible – and the same goes for cosmic rays. An alternative focusing device would be a static magnetic or electric field, though this would need to be strong enough to deflect the highly energetic cosmic ray particles to focus near to the centre of their collection area. There has been no observed evidence of such field effects in subjects while bending is in progress. Magnetic field measurements have shown no such change in such strengths near the specimen. Nor have electroscopes indicated the presence of any stationary electric field of appreciable strength. Fields far stronger than that naturally produced by the earth would in fact be necessary for such an effect to occur; and so we must rule out as well the idea of cosmic rays focused by magnetic fields.

As far as the specimen itself is concerned, the possibility of distortion arising from a differential temperature gradient can also be excluded. In order for this to cause the bending of an aluminium strip 10 centimetres long and 0.2 centimetre thick into a semi-circle, a temperature difference across the upper and lower surface of about 4,800°C would be needed! Even to obtain curvature of the specimen so that one of its ends bent to an angle of only 10° to the other, a variation of about 270°C between the two surfaces would be required. This is the kind of distortion one often sees in Gellerized specimens of that thickness and much thicker ones – which would need correspondingly higher temperatures. But no evidence for a simultaneous temperature rise has ever been observed in specimens which have been Gellerized. So we can also rule out any thermal mechanism.

However, there could still be localized heating which may have caused breakage of metallic objects by softening them to near-melting point. Analysis was made of a silver-plated spoon which had been very rapidly broken by Geller. No evidence could be seen of any change of the silver plating near the fracture, so it was concluded that an increase of temperature of no more than 200°C could have occurred at the fracture surface. To obtain enough softening for the fracture to have been initiated by partial melting would have required a temperature increase of about 500°C. Further evidence that local heating is not causing the bending or breaking is provided by examination of the imperfections which account for the mechanical properties of the metal, such as shearing and fracture properties.

Since this question of the structure of metals is of considerable importance to our discussion, we will enlarge on it. Metals are composed of a reasonably rigid lattice of atoms which have each lost an electron. The resulting positively charged particles, called ions, line up in a regular lattice held together by the negatively charged electrons flowing freely between them. The most important defect in such a structure is the presence of an extra set of ions; this imperfection is called a dislocation.

It is possible to understand how a metal bends in terms of the movement of these dislocations through the metal when pressure is applied. The region of extra ions travels through the metal much more easily than if one portion of the metal slid bodily past the rest. New dislocations are also created when pressure is used, allowing more distortion of the metal. They tend to get in each other's way, so the metal starts to become more rigid. Finally there will be so many dislocations that they will begin to join up, producing a small crack. This can then grow till fracture ultimately occurs. There are basically two types of resulting fracture: brittle fracture, in which the break is quite clean, and ductile fracture, where there is considerable plastic flow of the metal before it separates into two parts.

The structure of a piece of metal is not usually as uniform as described above, such regularity occurring only in crystals.

Metals are normally composed of many small crystals called grains, one next to another. If a metal is heated near to its melting point the grains begin to fuse together and so become considerably larger. At the same time the number of dislocations is reduced, so that they no longer get in each other's way. The metal thus becomes softer; it is said to have been annealed.

Returning now to the silver spoon which Geller broke, microphotographic pictures of its surface at different places, both close by and away from the fracture, give no evidence of any effect corresponding to a rise in temperature of 500°, high enough for annealing to happen. The fracture in this and other similar cases appears identical to one produced by bending the spoon backwards and forwards until it broke.

This does not preclude localized heating as a cause of fracture in other specimens. There is one aluminium strip which a subject distorted so as to have bends of reasonable curvature of several centimetres at either end, together with a very sharp fracture near its mid-point. Here, two different mechanisms could well be at work, one of a more general form (producing the smooth bends) and the other far more localized (generating the sharp fracture).

The feature of highly confined deformation of the grains has been discovered in other specimens. The first case was in a strip of thin copper, only 0.1 millimetre thick, which a subject had caused to break with softening occurring in the process. The fractured strip was then electro-polished and examined under high-power magnification. No evidence of deformation of the grains was found to within two grain sizes from the fracture surface; at some points the break occurred with no hint of deformation of the grains next to it. Grain distortion is expected at a distance of at least the thickness of the specimen away from the fracture's surface. In this case that would correspond to at least ten grain sizes. The only way that such a fracture could be obtained in the laboratory was by tearing the copper strip rapidly.

A second case of this type was later discovered in a strip of copper twenty times thicker. It was now impossible to conceive of the specimen being torn in two except by a very strong force indeed. Yet all that was needed to produce this particular fracture was gentle stroking. This type of break is to be contrasted with the typical ductile fracture – as if frequent bending backwards and forwards had occurred – which appears in much of the Gellerized cutlery investigated. It is also to be compared with the brittle fracture – the snapping apart of the two halves with no flow of the metal before breakage – which has occurred in other specimens. So in this case several mechanisms may be in action.

One of these, which has been suggested earlier, was the action of chemicals on the surface of the specimen. There are various substances which have a highly corrosive action on metals. They can penetrate the surface and cause defects to be accentuated. This produces a weakening which allows the metal to be broken more easily. Experiments have been tried with various chemicals. But the action of those which were tried could not satisfactorily account for it. A stainless steel spoon, for example, which was treated for five minutes in a solution of metal halide needed to be bent vigorously three times before snapping. A brass key painted with the same liquid turned grey and snapped when it was bent.

Chemicals of this kind could probably be developed to reproduce the physical distortion of the specimens caused by direct contact, though obviously they can never be regarded as a substitute mechanism in the examples of bending at a distance. Even in those cases of bending by direct contact which have been performed by those old enough to have had access to such chemicals, there is absolutely no evidence of any chemical action on the specimens. There would be very clear traces of this on the bent objects – discoloration and characteristic fracture lines – and none have been found. Many of my subjects would have had no chance of obtaining such chemicals or even of knowing about them. Nor have I ever, in any of my work with subjects, including Geller himself, seen any trace of their use.

The chemicals which can attack metals are in fact extremely toxic to humans. One which has been tested, for example, is absorbed through the skin and destroys the kidneys; as little as two-tenths of a gramme are enough to kill a man; moreover the effects are cumulative. Great care has to be taken in handling such compounds. As one of the chemists (a teacher) who had tested one of them remarked, 'I refuse to handle it without rubber gloves, and I change to a new pair each time.' So we can dismiss this use of chemicals as impossible in our subjects aged between seven and fourteen, who would have had little or no knowledge of chemistry, and tellingly remained in good health during several months of testing.

Let us now turn our attention to the possibility of vibration as a factor in metal-bending. This seems a much more plausible mechanism. Vibrations in a bar or cylinder of metal can be set up either along or perpendicular to its length. These vibrations can only occur at certain natural frequencies or wavelengths. For example, a strip of steel of 10 centimetres in length, 0.5 centimetre in width and 0.2 centimetre in thickness, held rigidly at one end, has its lowest natural frequency of vibration equal to about 130 cycles a second, and the next one up at about 800 cycles a second. Other strips have their own characteristic frequencies of vibration of about the same order. If it were possible to set a metal strip vibrating at one of its natural frequencies, then energy could be fed into it which might ultimately cause the strip to distort. This phenomenon of resonance is well known to bridge builders. The Tacoma Narrows bridge in the state of Washington, USA, was destroyed because the designers failed

---

Right: Movement, seen from top to bottom, of a dislocation across a crystal to produce the effect of slip. The dislocation is seen as an extra plane of atoms in the second diagram, as having been produced by a force applied from the left. This extra plane moves further to the right in the third diagram when further force is applied, and finally comes out at the far end in the last diagram. This gives the effect of slippage of the top half of the crystal over the bottom half.

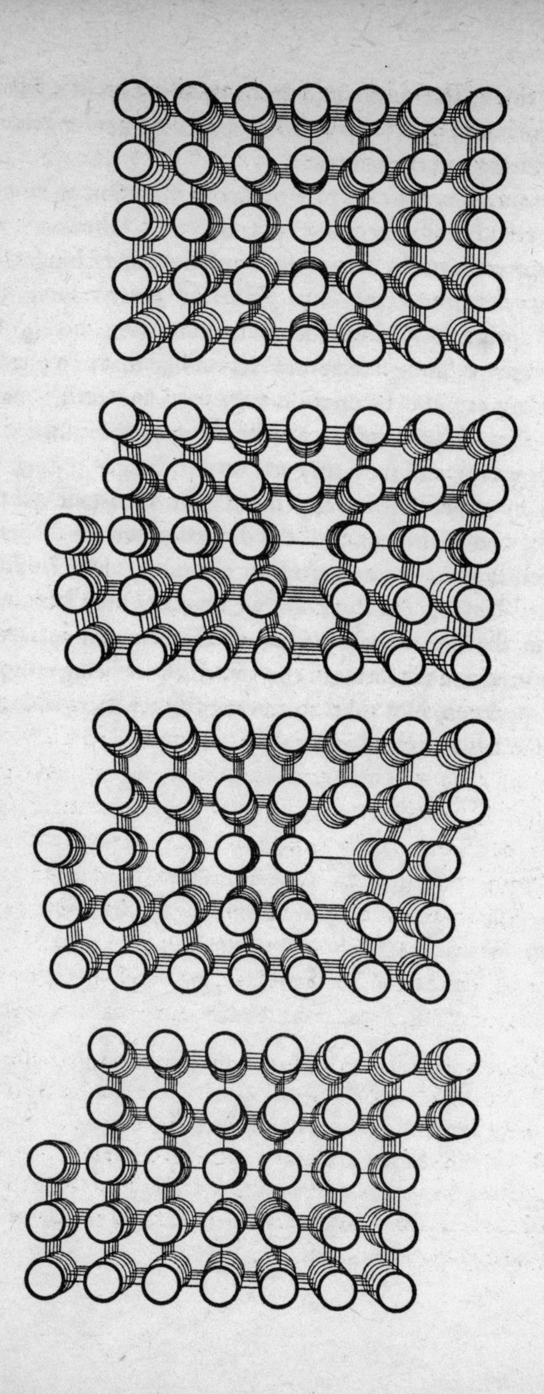

to take this characteristic into account. A group of soldiers are always required to break step when marching over a bridge for fear of setting it in resonance.

Resonance is not the only example of a vibration phenomenon which could cause bending and fracture. Ultrasonic waves (sound at frequencies above the usual human hearing level of 20,000 cycles per second) have proved of great value in metallurgical processes. Ultrasonic drills can form holes of any required shape, using a reciprocal action like that of a pneumatic drill. They can also be used on very hard and brittle materials such as glass and germanium. Metal fatigue can be caused in metal by such waves, and they are even used in fatigue testing. These and many other phenomena mean that ultrasonic vibrations could be one of the causes of the distortion we are discussing.

Evidently there are numerous mechanisms which could cause the metal-bending. At this point we must begin to look in more detail at the intentionality field, so that we can specify with greater precision the forms of energy which are being transmitted to the specimen. Not till then can we further narrow down the field of suitable mechanisms.

6 The forces of nature

The fields of force which we shall first discuss as possible candidates for the intentionality field are very much weaker than electromagnetism. We shall start with the weakest field in the universe, that of gravity. Its ability to hold us all on the earth depends on the presence of and the resulting pull of a great mass of matter. A comparison of the strength of gravity and that of electricity, for example, is given by the ratio of the gravitational attraction to the electrical repulsion between two electrons a centimetre apart.

This ratio of gravitational to electrical forces is $1:10^{43}$. Since 10^{43} is a huge number – 1 followed by forty-three noughts – this is a very small value indeed. On these grounds one is inclined to dismiss automatically any model of the intentionality field based on gravity. The weakness of gravity can be exemplified by the fact that the electrical charge on a single electron is still a million billion times more effective in repelling another electron a metre away, say, than is a human body in attracting that electron by the force of gravity at the same distance.

Only if a person could increase the strength of the gravitational effect produced by his body on others might gravity conceivably be involved in the intentionality field. There is a chance that this would result in such bizarre effects as the person rising from the ground and floating up to the ceiling. There have indeed been records of levitation – D. D. Home, for example, who was mentioned earlier. We might also find ways of explaining table-lifting in these terms, as well as the bending of metal and other objects, even when they happen at a distance.

One model we might suggest would require the modification of the basic mechanism of gravitational attraction. The keystone of modern gravitational theory is the fact that in a given gravitational field all bodies fall equally fast. A ball of lead and a feather fall to the earth at the same speed, providing they are in a vacuum so as to reduce differences caused by forces associated with the frictional resistance of the air. The theory that all bodies fall equally fast has been proved to be true for all the particles of which we humans are constituted – protons, neutrons and electrons – to at least one part in ten million. By changing this state of affairs, so that some particles fall faster than others, one might hope to cause levitation. But to do this so as to produce a repulsion between a subject and the earth would require the ability to change mass at will; there is no evidence at all for such mental control over gravity, other than that produced by the supposed acts of levitation.

An alternative is to alter the universal gravitational coupling constant, first introduced by Newton. In simple words, a massive body exerts a gravitational attraction on another body. This attraction is proportional to the product of the masses of the bodies and inversely proportional to the square of their distance apart; the constant of proportionality is Newton's constant, G. Modern theory does give a hint that modification of G could happen at very high energies, when the effective gravitational attraction between particles begins to increase. This change of gravity is only relevant, however, at energies and concentrations of matter that would be impossible to achieve in everyday experience. Indeed they have not yet been obtained on earth, nor is there any indication that such features occur in stars, though they may do so in black holes. The chances of small children achieving such conditions in their kitchen, or even of Uri Geller doing likewise on a theatre stage or under television lights, seem minimal.

The gravitational disturbances, caused when massive objects move violently, travel through space as gravitational radiation, in much the same way as a pulse of light is emitted from a torch. Direct observation of gravitational radiation has not yet been well authenticated. This is because the amounts of energy from distant stellar catastrophes, which generate the radiation, are extremely small. The level of power which could be radiated by a human subject as gravity waves is infinitesimal in comparison with that needed to cause metal-bending. In any case the person would have to jump up and down or twirl around to produce waves, and even then the energy output would be minuscule. Only if the subject dematerialized and then rematerialized (fast enough to avoid his absence being noticed) could any appreciable level of power output be achieved. There is no place for such a process in the framework of modern science; and there is no good evidence for its occurring. It seems reasonable, therefore, to dismiss gravity waves as a possible cause of the Geller effect, or of any other extrasensory phenomenon.

We turn, then, from gravity to the other 'weak' force of radioactivity. Radioactivity was discovered by the blackening of a photographic plate placed near a salt of uranium; the process was found to be due to the emission of three types of radiation, called alpha-, beta- and gamma-rays. The alpha-rays are composed of the nuclei of helium atoms travelling at high speeds. They cannot usually penetrate a few sheets of ordinary paper, let alone thin aluminium foil. Beta-rays were discovered to be composed of electrons, and their rays are more penetrating but are absorbed by sheets of aluminium several millimetres in thickness. Gamma-rays are the most penetrating form of electromagnetic radiation and can traverse aluminium plate several centimetres thick.

On the emission of alpha-, beta- and gamma-rays, the properties of nuclei change. In particular the electrical charge changes so that their chemical characteristics become modified. For example, the uranium atom decays through a sequence of emissions to end up as one of lead, with a total loss of eight alpha particles and six electrons; the whole process takes on average four and a half thousand million years. The only naturally occurring nuclei which undergo radioactive disintegration are the heaviest nuclei, but lighter ones may be made unstable by acquiring extra neutrons by neutron irradiation. These nuclei are the artificial radio-isotopes which have proved so useful in medicine as radioactive tracers and in radiotherapy.

We must now consider whether radioactivity could be a source of the intentionality field in the Geller effect and, just as with gravity, the large amount of energy required argues against it. There is very little indication of radioactivity in living things.

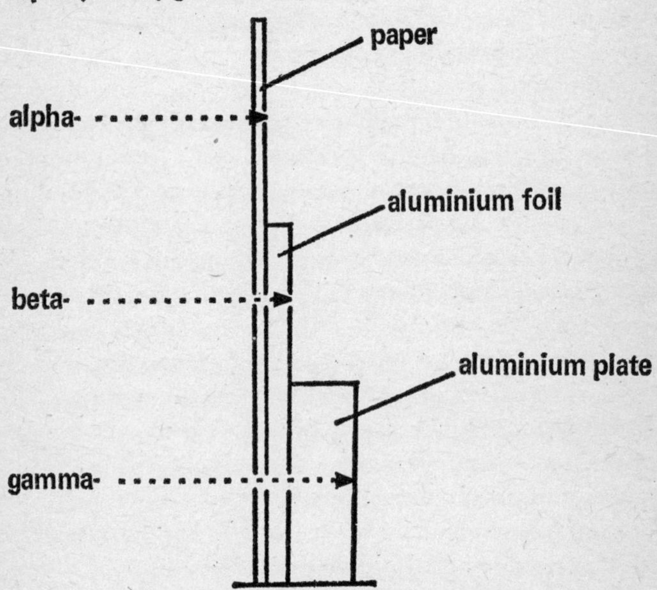

The penetrational power of alpha-, beta- and gamma-rays. Alpha-rays are composed of the nuclei of helium atoms and only penetrate a thin sheet of paper. Beta-rays, which are electrons, can get through aluminium foil, while gamma-rays can even penetrate a centimetre of aluminium plate

This is to be expected since the energetic alpha-, beta- and gamma-rays are all dangerous to health. Radiation can penetrate into cells and disrupt cellular processes, especially in the cell nucleus, where it causes genetic damage. Had the body of one ancestor contained natural radioactive material there would have been a very high mutation rate, which would have had the disastrous consequence of a very high frequency of abnormal offspring unlikely to survive.

Added to this overwhelming argument against radioactivity as a mechanism for the intentionality field is the fact that no energetic radiation has ever been detected while bending is occurring. My subjects have been tested using various radiation monitors which should have been able to detect such emissions, but no effects have ever been able to influence electroscopes directly, although they have tried hard enough to do so on a number of occasions.

Both the theoretical arguments against the involvement of radioactivity and these experimental results seem to be at variance with earlier results: Geller can himself affect both the Geiger counter and the Schmidt machine, and other subjects have been found who can influence either one or the other. But the complete absence of any effect on an electroscope rules out radioactivity *per se*. Since all other detection equipment apart from the electroscope operates using electronic circuits, only one hypothesis makes sense. It is that some electromagnetic field is disturbing the electronic circuits of these radiation detectors and causing them to simulate the occurrence of radioactivity.

Parents would of course be quite horrified if it were found that their metal-bending offspring were using these highly dangerous radiations to accomplish their feats. But the monitoring of many bending sessions shows that there is no risk whatever of this occurring. Let us also recall my earlier conjecture that the agent responsible for the Geller effect is, in fact, non-ionizing electromagnetic radiation. We will have to return to this idea in due course to see how it stands up to further analysis. In the meantime, we should take a look at other models which have recently been suggested.

Strong claims have been made that various strange particles are the underlying cause of ESP. The most impressive of these is the tachyon, a particle supposedly travelling faster than light; its name derives from the Greek for 'swift'. There is, it seems, no physical law preventing the existence of these particles, but it would be very disturbing if they were found, as they have very bizarre properties. For example, tachyons speed up as they lose energy, so that when they have no energy at all they are travelling at infinite speed. Tachyons can never travel as slow as the speed of light. Nor can any particle which at any time was moving slower than light ever breach the light barrier and turn into a tachyon. This is because it takes an infinite amount of energy for any particle, be it a tachyon or some other particle, to travel at the speed of light. The light barrier is impenetrable. Tardons (slower-than-light particles) and tachyons can never interchange roles. Let us suppose tachyons do exist and they can interact with tardons; if they could not, then we should dismiss them as being of no interest, since we would never know anything about them. This supposed interaction could theoretically allow a person equipped with a tachyon emitter, like a torch, to signal back into his own past. Such a process is clearly a violation of causality, as we pointed out earlier, and therefore totally at odds with the scientific viewpoint. Many serious attempts have been made to devise a scientific method of analysis appropriate to a universe where such signalling back into the past could occur. All these have failed on account of the absence of any natural limitation on the amount of acausality which could be achieved. But the past could be altered as much as one wanted if sufficient energy were expended on transmitting tachyons into the past. One might, for example, envisage a future in which fusion research had enabled large amounts of energy to be generated. If tachyons existed, this energy could be applied in building time machines, to travel into the past to modify it. This would annihilate both the fundamental tenet of science, that of causality, and also the regularity we see around us as expressed by its causal structure.

It has been claimed that the apparently coherent structure we see in the world only emerges because we have selected it out of the chaos of events which deluge our senses. This may well be true, though the success of science so far is therefore very surprising. It may now have met its match, however, in the Geller effect. To make progress we have to turn to experimentation. The evidence for tachyons' existence is almost nil. Theoretical considerations indicate that tachyons are most likely to be created in very energetic reactions, if they are created at all. Careful searches have been made of cosmic rays and in machines which accelerate sub-nuclear particles to high energy, but barely a trace of a tachyon has been found. Even the faint clues which have turned up do not lend much support to the idea that they could be the mediators of extrasensory phenomena. Limits have been put on the ratio of the force exerted between two bodies due to the exchange of tachyons and their gravitational attraction, which show that the former is at least a hundred billion billion times smaller than the latter. Since we have already shown that gravity can be discounted as the cause of the intentionality field, tachyons would be even worse.

There are other as yet undiscovered particles – intermediate-bosons, magnetic monopoles, quarks – popular among some investigators in ESP, which could be regarded as possible candidates for the Geller effect. The intermediate-boson is a particle essential for radioactive decay, and as such is likely to be as unsuccessful as the radioactive force it generates in furnishing a model for the intentionality field.

This is not quite the whole story, however, and a further comment needs to be made about the intermediate-boson. It has to do with the recent developments in physics to bring about the unification of the fundamental forces of nature. By means of very subtle arguments, the intermediate-boson has been combined with the photon to realize a unified theory of electromagnetism and radioactivity. According to this theory, effects caused by radioactivity could be much stronger than those normally observed; such increases could only arise, however, if intermediate-bosons are directly emitted. If discovered, the intermediate-boson is expected to be a very massive particle, at least

particle	*properties*
tachyon	Travels faster than light. Embarrassing if found
magnetic monopole	Magnetic equivalent of electrons. May be identified with intermediate-boson
intermediate-boson	Mediates the forces of radioactivity. May have been discovered already
quark	Conjectural constituent of protons and neutrons
psi particle	Newly discovered (November 1974), three times as heavy as a proton and long-lived. Other properties not yet known

forty times heavier than the proton, and having either a positive or negative electric charge or being neutral. Owing to its mass, it may only be produced in extremely energetic reactions, which may be the reason why it has not yet been created on earth. It could only be implicated in the Geller effect if very energetic protons, neutrons or electrons were generated by the subject; these particles could then create intermediate-bosons. But the quantities of energy needed are so great as to be quite impracticable.

One could conceive of some sort of accelerating action being achieved by the electrically excitable nerve cells in the brain. A proton or electron might acquire energy by a sequence of jumps, say from one cell to the next, and so on. Even if the surrounding material did not hamper the motion of the accelerated particle, the same difficulties raised earlier in respect of alpha-, beta- and gamma-rays would still apply. Such radiation, if produced, would cause deflection of an electroscope, and would also be genetically harmful to anyone emitting it. Still more to the point, the emission of radiation of sufficient power to cause metal-bending would burn holes in the skull. It would also affect behaviour patterns by acting on brain tissue. So radioactivity will not do as a model.

The neutrino, mentioned earlier as the electrically neutral, massless particle usually released in radioactive decays, has also been suggested as a means of achieving telepathy. Might it also account for the Geller effect? Since neutrinos have no electric charge and do not interact with any particles in the nucleus, they have great penetrative power. But this feature presents great difficulties regarding both the generation of a large enough signal and its reception. The only way such transmissions can be achieved is to use very high energy neutrinos: for then it turns out that the coupling with normal matter is increased. Here again, there seems to be no method of achieving such high energies by using the brain, nor, if they were to be achieved, of avoiding the ensuing tissue damage.

Magnetic monopoles – the unit magnetic north or south poles analogous to the electron and positron as units of electricity – must also be considered. The monopole has been searched for extensively, but no trace of it found. Recent suggestions show it can be identified with the neutral companion of the photon in the unified theories of radioactive and electromagnetic interactions. This is a beautiful idea, but as a candidate for the effect it has the same problem facing it as the intermediate-boson. For both from theory and experiment the monopole must have a mass at least five times as heavy as the proton, if not more. Only very energetic protons, neutrons or electrons could produce the monopole, probably killing the subject in the process.

The quark is the remaining hypothetical particle to be considered. It has been proposed as the constituent of the proton, neutron and their associated excited states; extensive searches have likewise produced no evidence for the existence of the quark. There are strong theoretical reasons why there should be

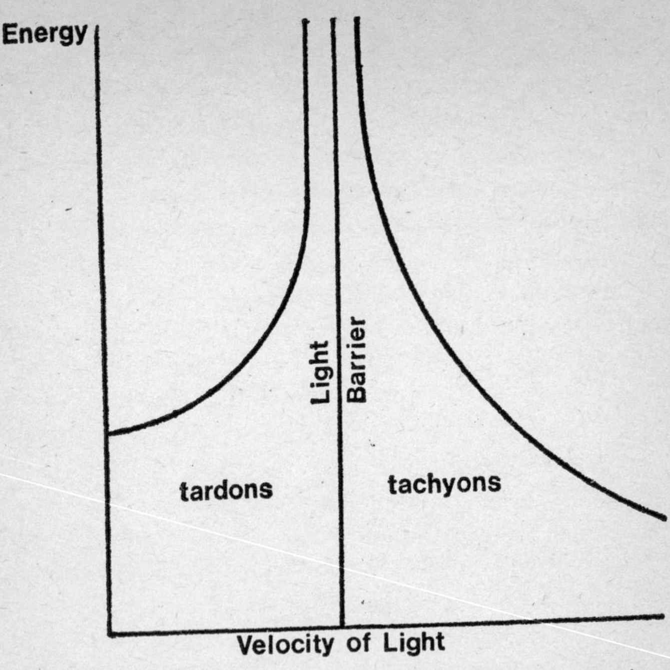

The variation of energy with velocity in particles which are slower than light (tardons) or faster than light (tachyons). Neither can broach the light barrier

quarks, but *if* they exist, they must also be very heavy, at least five times heavier than the proton. They therefore fail for the same reasons that monopoles and intermediate-bosons do: the particles needed to produce them are at impossibly high energies.

The same reason applies to beams of protons, neutrons or electrons. They exist in copious amounts in the body, but for protons or electrons to be penetrating enough to escape from brain tissue through the skull and surrounding skin, they must have energies of at least a million electron volts. Again, they should have affected the electroscope, but as no effect has been observed their involvement can be excluded. There was also no evidence of ill health among the subjects, which there would almost certainly have been had these types of radiation been produced. Neutrons can penetrate matter much more easily than electrons and protons, as they have no electric charge causing them to be repelled by atoms of the material through which they are passing. Yet there seems to be no mechanism in the brain for accelerating them: they too can be dismissed.

7 Electromagnetism and life

Some people have always had a strong desire to control the minds of others. This has led them to dabble in the occult, and even to claim to have acquired such control. The first man to formulate any theory about the power of mind over mind was Franz Mesmer, in the latter half of the eighteenth century. He believed that a diseased body could be cured by stroking it with a magnet, and that effects on others could be obtained by 'animal' magnetism. The ideas excited the public fancy so much that ladies in his home city of Vienna wore 'magnetized' clothes and ate from 'magnetized' plates. Mesmer invented a wooden tub containing magnetized iron filings, from which protruded flexible metal rods for the sick to touch. He was later investigated by a commission appointed by Louis XVI in Paris. They issued a report attacking his idea of animal magnetism, suggesting that it might disrupt public morality. Society withdrew its patronage and mesmerism became disreputable. The undeniably effective factor of hypnotism which he employed in making cures has only recently become recognized as a legitimate tool in healing by medical professions in some countries.

Animal magnetism itself has suffered equally. For a long time very little scientific analysis of the idea was carried out. Only quite recently have any searching experiments been performed to detect the effects of magnetism, and of electricity, on humans and on other living forms. This is still only one half of Mesmer's original idea, since it does not take into account the possible fields generated by the brain or by living tissue. This other aspect cannot be neglected in our discussion of the Geller effect; we will return to it later.

In recent years a great deal has been discovered about the action of electricity and magnetism on living tissue. A large range of human ailments have been found to have been caused by excessive electrical or magnetic activity. On the other hand, modified forms of this activity have been discovered to be remarkably beneficial both to humans and animals. Let us run through what is known about this before considering its relevance to the Geller effect.

The first phenomena to consider are those caused by naturally arising electromagnetic activity – electric storms, various forms of dry winds, variations of the earth's magnetic field, sunspot activity and so on. There is no doubt that these processes have their effects on humans. For example, a clear-cut relation has been found between the occurrence of electric storms and thrombosis. Even more extreme effects are produced by warm dry winds, such as the one which moans its way from the Russian Steppes to the Alps. The effects of such a wind are varied but quite strong; even today surgeons in many countries try to avoid operating when the wind is at its height owing to a greater frequency of post-operative complications thought to be concurrent with it.

These influences of the weather on human health have long been appreciated; almost 2,400 years ago, Hippocrates wrote: 'North winds bring coughs, sore throats, constipation, retention of urine . . .' There is a wind called the Sharav in Israel, which upsets about thirty per cent of the population. It blows for about one-third of the time in Jerusalem, and rather less often in the cooler, moister coastal plain. Yet there is as much of certain kinds of sickness during the days the Sharav blows in either area, in spite of their different humidities and temperatures. The reactions to this ill wind include a good proportion of human so-called 'psychosomatic' ills: sleeplessness, irritability, migraine, allergies, nausea, vomiting, flushes, sweating, fatigue, apathy, exhaustion, depression, confusion, ataxia, palpitations, and other similar conditions. These symptoms have recently been found to arise from modification of the action of the hypothalamus, part of the brain, of the thyroid or of the adrenal glands. There are, incidentally, a number of famous men who have suffered from hyperthyroidism, notably Goethe and possibly Mozart.

The root cause of these reactions, as well as of the increased incidence of thrombosis during electrical storms, has been found to be the greatly increased number of positively charged particles in the air. Those particles, called positive ions, arise from the loss of electrons from oxygen or water molecules in the air by the friction of the wind or by electrical activity. Such positively charged particles can congregate on the surface of a living cell and change its metabolic activity; it is thought that this produces the symptoms listed above. The effects may be felt even two or three days ahead of the arrival of the wind itself, as is the case with the Sharav, coinciding with the increase in the number of positive ions in the air. The density of these ions during the Sharav may rise as high as 4,500 per cubic centimetre, three times that in normal conditions.

A simple method has been found to cope with this unpleasant effect. It is to generate negative ions with a suitable machine, and numbers of these are now commercially available. These machines have also been used in American space capsules and in long-range reconnaissance aircraft. It is interesting to note that running water can produce beneficial negative ions; as the physicist Philip Lenard wrote in 1892, 'It has been known for a long time that waterfalls will charge the ambient air with negative electricity.' A slight negative electric field results from flushing a toilet with the lid raised, and successively stronger negative fields from running water into a hand basin and into a bath. The most satisfying field of all, of about 1,000 volts per metre, can be produced by turning on a shower.

Atmospheric ions have effects on many other organisms besides man. Bacteria show increased mortality rates at high ion concentration in a smog-containing atmosphere; hamsters show less tolerance to carbon dioxide in air containing positive ions as compared to that containing negative ions. The moulting of aphids is increased when they are exposed to air containing negative ions and is drastically reduced when a high concentration of positive ions is present.

Naturally occurring electrical fields have great survival value for some species. The electric eel, for example, has eyes which are useless for navigation; it finds its direction by means of an underwater electrical radar system. The West African fish gymnarchus can swim equally well backwards or forwards by means of the electrical impulses generated near its tail. A class of fish, the elasmobranchs, which embraces sharks, rays and dogfish, can detect the minute electric fields of one millivolt per centimetre generated by the gill movements of the plaice, even though the latter may have buried itself in the sand. The elasmobranchs have electrical sense organs (exotically called the ampullae of Lorenzini) seated in deep skin pores around the snout, and these are sensitive to electrical fields ten thousand times smaller even than those generated by plaice.

Magnetism also occurs naturally, the earth's magnetic field having a value of about a quarter to half a gauss (the unit of intensity of a magnetic field), with variations in direction from point to point on the earth's surface. One might expect that such a field will have played a role in animal evolution, and that animals will have developed cells sensitive to the direction or strength of the magnetic field. And this has been found in homing pigeons. When magnets are attached to them, so destroying the earth's field, they cannot return to their loft on an overcast day (when they cannot use the sun to navigate). *How* the pigeon detects the direction of the earth's magnetic field is unknown, as in the case of the many other organisms, like mud-snails or planaria, which can do the same.

Humans are also sensitive to the magnetic field. During periods of relatively large fluctuations in the earth's magnetic field, admission rates to hospitals have been known to rise, and over a period of five years a striking correlation has been discovered between illnesses and deaths and a series of sharp terrestrial magnetic disturbances. Another year-long study has disclosed similar variation between geological parameters and the fluctuations of the earth's magnetic field.

It has been claimed that dowsers are also sensitive in this way, even to very small disturbances in the local magnetic field. There is some experimental support for this; in laboratory tests dowsers can detect if a current is flowing in a large coil of wire past which they are walking. Various mechanisms have been postulated in support of this, but neither the experimental nor theoretical situations are at all well defined. The most careful experiments have been made by Professor Rocard of Paris, who claims to have shown that a good dowser can accurately detect a variation of about three ten-thousandths of a gauss per second. This alteration of magnetic field is usually produced when the dowser walks at a normal speed past local variations of the earth's magnetic field of a few ten-thousandths of a gauss taking place in a few seconds. This magnetic anomaly, as it is called, is thought to be produced by underground water filtering through a porous medium or water lying in permeable layers next to beds of clay. In both these cases, electric currents could be produced which would then generate a small magnetic field at the soil's surface. Rocard has also suggested that variations of the magnetic field cause nuclei of molecules in the blood to rotate at different frequencies from those in the bones; the 'beats' of a few cycles a second thus set up could well have a large enough effect to be detected by a physiological response such as a muscle twitch, as in the case of a dowser. But this explanation doesn't tell us how dowsing can be successfully accomplished when the operator is working only with a map of the area.

Living forms which have become adapted over the millennia to a certain magnetic field may suffer when it is removed. This problem is very relevant to interplanetary travel, where extremely low magnetic fields have been discovered. For example, a magnetic field of about a hundred-thousandth of that on the earth's surface would be encountered on a voyage to one of the other solar planets. To prepare for such an environment, man and other animals have been exposed to low fields in suitably shielded enclosures. The responses observed in mice kept for a year in a magnetic field of about a thousandth that of the earth range from a shortened life span to infertility and cannibalism.

The only effect on men kept in such an environment for ten days is a reduction in their ability to recognize that a flickering light is actually flickering and not continuous. Simple algae are found to have their growth rate accelerated by such fields, while bacteria show a fifteenfold reduction in the size and number of their colonies. Clover seeds increase their rate of germination in a low magnetic field, while the direction the roots of winter wheat grow in can be determined in an otherwise magnetically shielded environment by the direction of an applied field of about the strength of the earth's.

When a variable electric field occurs it always has a companion magnetic field arising from the electrical variation. Similarly, a varying magnetic field is accompanied by an electric field, so that the two are considered together. Electromagnetic radiation occurs naturally in many wavelengths, some of which have been described already – in order of increasing frequency and decreasing wavelength, these are: low frequency, radio waves, microwaves, infra-red, visible light, ultra-violet, X-rays and, with the shortest wavelength of all, gamma-rays.

Sensitivity to environmental cues such as electromagnetic radiation has evident survival value: eyes have developed in large numbers of animals. Some animals can also detect infra-red radiation. Pit vipers possess organs on their heads which give them stereoscopic infra-red vision, so enabling them to see and strike accurately at their prey in the dark. All living things emit infra-red radiation by virtue of their temperature – infra-red rays could be called 'heat waves'. It is possible, by using special cameras, to photograph the heat rays emitted by the body. The resulting thermographs are of great value in diagnosing, for example, cancer, where tumours are identified as regions of higher temperature.

Some people claim to be able to see an aura radiating from others. The fact that saints and the hierarchy of heaven were depicted with haloes may have something to do with this. So far there is no way of objectively measuring the human aura. But perhaps it will become possible by determining the infra-red aura of the body. The human aura is supposed to extend far less in sickness than in health, appearing to 'droop' according to those who can see it. The general features of the aura seem very comparable to pictures obtained by the thermographic camera. Since about two to three per cent of people can see an appreciable way into the infra-red spectrum (a hot iron will glow for them in a dark room) then it is a reasonable conjecture that the aura is part of the infra-red radiation from the body. This idea can be tested by measuring the infra-red sensitivity of those who claim to see the aura.

The body also radiates in the ultra-violet region, though to a much lesser extent. There is no strong evidence that the human aura is in this region, though numerous arguments have been put forward in favour of it. Diacynin dye and similar substances have been claimed to render the eye sensitive to the ultra-violet spectrum by looking through a screen filled with the substance for a few minutes and to allow the human aura to be observed. But there is little substance to this, since the cornea of the eye is normally opaque to ultra-violet, and no way is known of rendering the eye transparent to ultra-violet other than by actually removing the cornea. The primitive ultra-violet sensors I used on Geller and my other subjects showed no response at all to any human body; the level of ultra-violet emission is certainly below that of infra-red, which is relatively easy to detect at a distance by a suitable thermometer.

The scientific revolution has allowed man to surround himself by devices whose dangers he has not immediately appreciated, such as the hazards of microwave ovens and X-rays from colour TV screens. Perhaps other parts of the electromagnetic spectrum can also endanger our health. Let us look at the effects of man-made radiation, starting with static electric and magnetic fields.

Since the body is a poor electrical conductor it is impossible to obtain effective static electrical activity in the deep organs without strong electrical current at the surface of the skin, probably destroying tissue in the process. Damage to living tissue does not occur to the same extent with magnetism, and magnetic fields of up to a quarter of a million times that of the earth have been used on organisms. Naturally enough, all sorts of reactions take place, including orienting and increased motor activity. The oestrus cycle in mice is abolished by a field of 4,000 gauss, while cancer patients exposed to fields of several thousand gauss have all experienced a reduction of pain and prolongation of life. Lower field strengths have been used to treat various complaints with apparent success.

Russian scientists have discovered that the action of a magnetic field above a hundred gauss on the animal brain slows down the firing of the nerve cells and increases the amount of slow waves of coherent activity over the brain which usually occur in sleep. The same happens in reptiles, pigeons, rabbits, monkeys, and in man. The response of nerve cells to light or sound stimuli is also inhibited. The mechanism underlying this effect was thought by the Russians to involve the consumption of oxygen by the brain, especially when a group of mice showed themselves less able to survive oxygen starvation when under a strong magnetic field than without it.

The effects of low-frequency electric current have been found to be of great value in treating arthritis, sciatica, shingles, sprains and strains, fractures and various other complaints. Various currents are fed into the body through electrodes. To avoid burning sensations due to too high a current, frequencies of about 4,000 cycles per second are used, for then the resistance of the skin to the passage of the electrical current is reduced. The skin, for example, obstructs the passage of a current with a frequency of 50 cycles about one hundred times more than one at 4,000 cycles. So with the less painful frequency, reasonable amounts of electric current can be delivered to deep organs in the body. Two of these alternating currents are set up so that they cross over at the particular site of the disease. If the frequencies of the two currents differ by, say, a hundred cycles, then they will interfere to produce a beat rhythm of frequency equal to this difference. The crucial feature of the treatment is the low frequency achieved by this interference of the crossed currents. Muscle fibres appear to be most successfully stimulated at interference frequencies of 0 to 10 cycles, whilst increases in general muscle and nerve tone and in blood circulation, as well as in sensory stimulation and relief of pain, occur at 90 to 100 cycles. These frequencies are roughly in the range of the natural rhythms which occur in our bodily processes.

Strong interest has been recently aroused in a method of photographing the surface of living tissue using high-speed electrons. Kirlian photography is named after the two Russian scientists who developed it in 1939. In the last few years other scientists have begun investigating the process and adding to the technical refinements. Basically, it drags the electrons out of the skin by applying a high voltage to it of between 20,000 and 50,000 volts. The electric field is applied in a pulsed or alternating fashion so as to reduce the resistance of the skin to electricity; frequencies of twenty thousand to three million cycles a second have been used by the experimenters. The electrons are accelerated to a velocity high enough to cause the emission of light, which is then detected on a photographic plate. The technique has been used most effectively to investigate the human skin. Kirlian photographs have been taken of the finger-tips of people in various psychological states: relaxed, tired, inebriated, suffering with a cold, hypnotized, in a trance; psychic healers, too, have been investigated. The Kirlian photograph shows flares or rays of electrons proceeding from certain points on the skin and varying in length or number according to the state of the subject. There appears to be both change in the length of the rays (they decrease when the subject is tired, and flare out if injury occurs to the finger) and in coloration – red blobs appear near the centre of the finger-tip just after a faith-healing session.

These effects are to be expected on general grounds from the change of the resistance of the skin to electricity in altering the state of consciousness: the resistance increases as the subject becomes more relaxed. This skin response has been used as a measure of the depth of a hypnotic trance, or, at the other extreme, as a lie detector, to observe when a subject is tense whilst answering a question. Some Russian scientists have gone much further than this, and have claimed that the Kirlian method has revealed what has been called 'bio-plasmic energy, a definite system of ionized particles which both surrounds and interpenetrates the physical body'. There is no evidence for this outside the Kirlian effect, but undoubtedly Kirlian photography is a new method of obtaining information about the physical and emotional state of a living organism.

Radio waves can in some instances be detected by various people through the fillings in their teeth (acting as miniature receivers). They can also be 'heard' directly. The meteorite which fell on Barwell, England, on Christmas Eve in 1965, was 'audible' to hundreds of people as a buzzing or hissing sound as it travelled over England. This is a phenomenon which has often been reported over the years, and as the meteors in these cases were in a region with too little air to produce sound waves, the noise must have been produced by electromagnetic radiation. Experiments were then carried out in which many people, on being exposed to a very weak beam of radar waves, perceived what they described as hissing, buzzing, clicking or knocking noises, depending on the characteristics of the radar transmission. So it does seem that some people are able to hear radio waves directly. A possible explanation of this was outlined by a former

Postmaster General of the USA: 'People exposed to radar beams have experienced a hearing sensation which has been described as a buzzing sound. It is at present considered possible for this sound to be caused by radiation-induced forces which are acting on the skull or part of the skull or the middle-ear structure . . . The implications of the phenomenon can be quite far reaching. Thus, one might speculate about the development of microwave communications techniques which permit one to communicate with a person at some distance via a microwave beam, even though the person addressed has no electronic receiving equipment whatsoever.'

An important question raised by this effect is that of the detailed nature of the detection mechanism. There could be forces set up in the skull, as suggested by the Postmaster General, which then set off a reaction in the appropriate nerve cells in the auditory area of the brain. The noises of hissing, clicking and so on are very close to those evoked by direct electrical stimulation of the brain. Support has come for this idea with the recent discovery that irradiation of an isolated turtle heart by microwave radiation of a certain frequency causes a decrease in heart rate. This only occurs over a very narrow range of incident power, whereas at higher power levels the heart rate is increased by a heating effect. It was further shown that the reduction of heart-rate came about because of the increased stimulation of remnants of certain nerve cells still sticking to the outer surface of the turtle heart. Microwaves, therefore, can stimulate nerve cells directly.

All in all, it is certain that man is sensitive to electromagnetic radiation over the whole range of frequencies – static, radio, microwave, infra-red, visible, and beyond. For radio frequencies we know his sensitivity appears to be high, and he can detect far less than millionths of a watt per square centimetre. One organ directly sensitive to the radiation may be the brain, while for infra-red and visible light his skin and eyes are good detectors. The organs of his body also respond to the therapeutic effects of low-frequency currents.

Knowing this much of electromagnetic radiation and its effect, we now turn to what influence one person could exert on another, or on an inanimate object, by means of electromagnetism. In other words, we must investigate the electromagnetic aura. Some aspects of electromagnetic emission from the body have already been discussed, in particular the infra-red and ultra-violet. The total power emitted in these frequency ranges is still comparatively small, totalling no more than that expected to be emitted by a man-sized object heated to about 98.4°F – less than 100 watts (compare this with a kilowatt electric fire). In any case this power could only be *radiated* to give heat.

The X-ray or gamma-ray aura, we saw, must be very small to allow the tissue of the body to survive; as must be that involving neutrinos, neutrons, protons, electrons, quarks and other particles. There is, in any event, far too little energy to produce these particles in any measurable amount. Only at the low-frequency or radio-frequency end of the electromagnetic spectrum could any appreciable aura be expected. And to this we turn.

The basic organ involved in behaviour is the central nervous system. This is composed of an aggregate of nerve cells or neurons, there being about a hundred billion in a human being. Each of these is electrically excitable in the sense that a large enough stimulus, such as an electrical pulse applied by an external electrode, will cause the internal potential of the cell to increase by about a tenth of a volt, and then return to its resting value, all in less than a thousandth of a second. This nerve-cell excitation is an all-or-nothing response and only occurs if the stimulus is above a certain threshold, but then it cannot be prevented by any normal means. The sudden change of potential, called the action potential, occurs first at a site on the surface of the cell body, and then travels uniformly down the cell axon. The axon is a projection of the cell body which may extend less than a millimetre, as in the brain, or more than a metre, as do axons projecting to muscles in the arm or hand from spinal neurons.

In the brain, axons can ramify considerably and usually end either on other cell bodies or on projections from them, called dendrites. Where the axon of one nerve cell joins the cell body or dendrite of the next is a gap between the two cells: it is called the synapse. The information that a nerve cell has responded in its all-or-nothing way is carried down the axon and its ramifications as the action potential; at a synapse it is converted into the transmission of a certain chemical across to the post-synaptic cell membrane. This chemical causes a change in the electrical potential of this latter membrane; if the change is large enough an action potential will be initiated. There are brain cells whose axons carry signals which go out to the periphery of the organism, carrying central commands to muscles of various types. There are also neurons whose one function is to transmit information about external stimuli to the brain so that it can take appropriate action.

Electrical activity also occurs in muscle fibres. These are specialized cells which receive information but give none out. They can be anything from one millimetre to ten centimetres in length, and each one is caused to contract by a nerve impulse carried down the nerve cell ending on it. Each such motor nerve cell will divide and innervate about one hundred to two hundred muscle fibres. The contraction of the latter is a result of the rapid depolarization of the muscle cell surface – by about a tenth of a volt, as already noted for the nerve cell, and again occurring in a thousandth of a second or less.

The electrical activity in the central nervous system and in the musculature will evidently contribute to the electromagnetic aura. The range of extension of this aura will then be partially governed by the level of activity occurring at the time. The frequencies making up the aura will be determined by those involved in nerve-cell or muscle-fibre activity. There is the problem that flesh and bone do not transmit all frequencies equally, and this should be taken into account when assessing the size and strength of the aura.

There are other contributions to the electromagnetic aura apart from the electrical activity in muscle fibres and excitable nerve cells. The surface of the skin itself does not remain at a constant electric potential. Different parts of the body may be at different potentials, which vary by up to a tenth of a volt. Superimposed on these steady differences are changes of potential, which occur either in response to an external stimulus or spontaneously. The response has three phases; the potential is first reduced by up to five-thousandths of a volt over about half a second. Then it is increased by a hundredth of a volt, only to return to its original value two or three seconds later.

The skin potential response and concomitant resistance changes can be directly measured by applying suitable electrodes to the surface of the skin. It is also possible to detect potential changes in the skin arising from brain, muscle or heart activity by using surface electrodes. The first of these involves electro-encephalography, electrodes being applied directly to the scalp to measure the microvolt changes (one microvolt = one millionth of a volt) of potential brought about by underlying neural activity. Most of the electrical information appears to occur in the range of one half to fifty cycles per second. Certain values inside this band of frequencies are correlated with behavioural activities, and have been isolated and identified.

One of the best known of these is the alpha rhythm, much prized by meditators for its predominance in states of relaxation. This rhythm occurs at about 10 cycles a second with amplitudes up to 50 microvolts. It occurs just before sleep, and can also be obtained by training in meditation or by recent electronic bio-feedback devices. In states of extreme emotion, theta waves of four to seven cycles and up to thirty microvolts occur. These are seen most frequently in children, their amount reducing as a person becomes older. There are also delta waves of one half to three cycles and up to a thousand microvolts (a thousandth of a volt) in amplitude. These slow waves recur regularly during sleep: hence the stage named slow-wave sleep. Finally, there are faster beta waves of about 20 cycles and up to 30 microvolts in size; these normally occur during waking activity.

Very rapid oscillations of between five hundred and a thousand cycles are also seen to occur with amplitudes up to two hundred microvolts. All these oscillations can occur in exceptional cases with up to twenty times their normal states. These brain waves are thought to be caused by the synchronized activity of many neurons. The higher amplitudes will naturally occur when larger numbers of neurons are acting in concert. Since it is to be expected that behavioural responses are determined by the patterns of brain activity, these latter have been used as an indicator of the 'mental state' of the animal. However, it is necessary to measure the brain-wave activity separately over different areas of the brain to define better the whole brain's activity. Even then it is not clear that the localized patterns of surface brain activity will ever produce a fine enough indicator of underlying brain activity.

It is only through detailed analysis of single-cell activity that certain aspects of brain activity can be discovered. Processing of visual information, for example, has been found to be the function of single nerve cells responsive to very specific portions of the visual field. Some cells will respond best to a slit of light at a certain orientation or to a spot at a certain distance from the eye; their response to slits at other orientations, above, below or at the side, or at other distances may be markedly reduced. These cells therefore act as pattern recognition units. The visual field is broken down into small spots by the nerve cells in the eye, and these are filtered through cells which recognize successively more complicated patterns. It may be that there are even recognition units for, say, specific faces. Such information processing and storage could never be discovered in gross brain-wave patterns and their changes with time. Yet the electro-encephalograph is still a very important indicator of general behavioural patterns, especially since it doesn't involve the major surgery of inserting electrodes into the brain and metering the output.

Changes of skin potential resulting from heart activity can be measured by placing electrodes at strategically selected positions on the body, for example on the arms; the electrical signals being detected there are generated by heart activity and conducted through body fluids to the body surface. The resulting signal, the electrocardiogram, has a number of components, the greatest being a change of about two thousandths of a volt due to the excitation of the ventricles of the heart. It is rare to find individuals with mean heart rates below 50 beats per minute or above 150 beats per minute; shifts can occur within a few minutes of 10 to 20 beats per minute. There is little heart activity at frequencies higher than 5 cycles a second.

Muscular activity can also be monitored by skin potentials. The resulting frequency range makes very little indication at less than 20 cycles a second, so that heart and muscle activity can be separated out in the detection apparatus. The signal detected by electrodes placed on the surface of a muscle depends on the number of muscle fibres in the vicinity of the electrodes, in other words on their size. If the total activity from a muscle is being measured there may be a potential change of up to a hundredth of a volt, due to the total output of potentials from separate motor cells. The frequency range of this activity is at its most effective between 100 and 600 cycles, with the maximum occurring at around 500 cycles. Much smaller levels of activity can be detected down to 20 cycles and even up to 5,000 cycles, the higher frequency range arising from the activity of a single muscle cell.

Heart, brain and muscle activity thus give rise to changes of electrical potential on the surface of the skin of no more than a tenth of a volt, with frequencies in the range of 1 to 5 cycles, 0.5 to 30 cycles (exceptionally up to 1,000 cycles), and 20 to 5,000 cycles, respectively. These effects are clearly relevant to bending metal by touching it, but to consider how they might be used to bend metal at a distance we shall have to evaluate how far these potentials extend their influence beyond the body to form the low-frequency part of the electromagnetic aura.

Very little experimental work has ever been published on this, especially in association with paranormal phenomena. One of the basic reasons is the great difficulty there is in detecting signals with a very long wavelength using a much shorter aerial. For example, electromagnetic radiation with a frequency of 3 cycles a second has a wavelength of a hundred thousand kilometres, and so will find it difficult to excite an aerial a few metres in size. Nor will an animal like man be expected to emit much radiation at such wavelengths, since the power generated by an aerial is proportional to the square of the ratio of the size of the aerial to that of the wavelength of the radiation emitted. Man's radiation is obviously weak. In spite of this there have been attempts to detect this radiation. The Italian neurologist Cazzamalli claimed in 1925 that he had detected emission from humans inside a lead-shielded chamber, with a wavelength between one and a hundred metres. Such measurements of animal electricity at a distance go back to the observations of Davy and Faraday on the discharge of the torpedo fish's electrical organ. But more recent work has not confirmed the work of Cazzamalli: it is thought that his results were due to a special coupling between the human body and the receiving aerial for the wavelength of about three metres.

Some Russian work has been done, using remote sensors which pick up the modifications in water molecules when they are transmitting electromagnetic radiation. Fluctuations of electrostatic and magnetic fields at 3 to 6 cycles a second have apparently been recorded. One particularly important result was of fluctuations greater than those measured by electrodes attached to the surface of the scalp. This was the case of Nelya Kulagina, the Russian woman who has psychokinetic powers. These values were said to have occurred when she was performing movement of distant objects. This result is obviously of great importance to the electromagnetic explanation of extrasensory phenomena, but needs authentication, especially as there is a lack of any precise information about the details of the experiment. High-frequency radiation (up to 150,000 cycles a second) of low intensity was said to have been detected some distance from her muscles during their contraction, but none during relaxation; this has since been repeated with frogs and humans. But only weak signals at much lower frequencies were registered. This low-frequency radiation is predictable, since it can be detected on human skin. Skin, tissue and bone are quite transparent to electromagnetic radiation of long wavelengths; less than half of it at 1,000 cycles, for example, is absorbed after passing through 35 metres of such body material, and there is even greater penetration of lower frequency radiation. Therefore the electrical activity of the brain, heart and muscles can easily be transmitted to the exterior of the body. The total power radiated in this fashion, however, is small. That emitted by the oscillations of charge on the skin, corresponding to the surface electrode measurements I described earlier, is less than a billion billionth of a watt! There may be considerably more radiation from deeper organs. For example, if a billion neurons were all active in a concerted manner at a thousand times per second, the total power radiated could be measured in micro-microwatts. This tiny amount is just measurable by modern techniques.

Many conjectures have been made in the past about the transfer of biological information from one part of an organism to another by means of electromagnetic fields. In particular, the Americans Burr and Northrop have suggested that in each organ is an internally generated electromagnetic field 'which is in part determined by its atomic physico-chemical components and which in part determines the behaviour and orientation of these components.' There is clear evidence, some of which has been described, for the electromagnetic field generated by the organism. But the other half of Burr and Northrop's hypothesis is not so well authenticated.

We have assembled knowledge about electromagnetism and the body and we can now apply it to the Geller phenomenon. There may not be enough data on hand yet to give a final answer, but at least we shall be able to see what are the crucial questions that still require answers.

8 The final explanation

The most satisfactory explanation contemporary science can offer for the telepathic and metal-bending powers of numbers of people, including Uri Geller, is that it is achieved by an electromagnetic field of force. The arguments given in Chapter 6 have eliminated all other possibilities: humans could not use radioactive, nuclear or gravitational forces to deliver enough power to a piece of metal or to another person's brain. The quantities of energy needed to activate the radiation corresponding to these forces is far too high to be considered at all likely.

It has therefore been suggested that the human operator emits the electromagnetic intentionality field which interacts with the receiver of a telepathic message or with a piece of metal. Now the electromagnetic explanation of extrasensory phenomena is not new, nor has it proved adequate in the fifty years or so of its existence.

From the points of view expressed in the previous chapter it is fairly evident that this explanation is not very reliable. Because the fluctuations brought about by muscular, heart or brain activity are of the order of thousandths of a volt only, the resulting variations of energy emission will be correspondingly rather small, far below the level needed to explain the bending of metal. Nor is it easy to understand how focusing of the radiated electromagnetic field is to be achieved if it has a frequency determined by that of the electrical activity of the body. This was seen to be below about 5,000 cycles per second, with a wavelength of at least 60 kilometres. Focusing over distances much less than that would be very difficult to achieve.

The obstacles these theoretical considerations present are formidable, and no doubt there will be support for the conclusion that the electromagnetic hypothesis cannot be right. Some people, and especially scientists, may even go on to deduce that because there is no rational explanation for these phenomena, they simply cannot exist. I cannot take that easy way out because I have personally witnessed the Geller effect and telepathy under conditions in which fraud can be completely ruled out. But if electromagnetism be eliminated as the cause of the intentionality field, what is left?

It was remarked earlier that scientific explanation must be attempted for these phenomena or science, and even reason, will have been found wanting. The possibility that there might be a fifth physical field of force has been discussed now and again, but from what has already been said, electromagnetism would seem to be more satisfactory. In addition, there is no evidence at all at the microscopic level for such a fifth force. Nor is there any hint from science of any force beyond the well-established four considered in Chapter 6. The only alternative candidate for the intentionality field is a so-called psi field, which is involved in some way with physical objects; it is not in *physical* space but in some other space or spaces. The notion of such a 'field' was introduced several years ago purely to explain some of the strange extrasensory phenomena discussed in this book. This completely non-physical character of the psi field is essential because otherwise there would be problems similar to those challenging the electromagnetic hypothesis.

There is, then, this problem to be faced concerning the psi field, namely that if it is entirely non-physical, it would never be able to interact with the physical world and cause the bending of metal. One would have to suppose that even though the psi field is not in physical space, and so has no difficulty in penetrating shielding or travelling long distances, it can still interact with physical objects so as to bend a spoon or somesuch. But this is a very complicated operation to postulate, with various spaces or levels and interchanges occurring between them. The non-physical spaces appear very difficult both to quantify and therefore to discuss in any way analogous to the model of our own physical world. One may conclude that if there are alternative candidates to electromagnetism as the intentionality field, they are even more unsatisfactory. One of these other possibilities is a completely physical and almost as completely unknown fifth force of nature; an alternative is a psi field which is only partially physical and just as unknown. Either of these two possibilities, however, leaves the Geller phenomenon as unexplained as before. The psi field has been introduced specifically to understand metal-bending and telepathy; unfortunately it remains completely unrelated to the other forces of nature, nor has the psi field hypothesis led to any predictions which could allow us either to confirm or to falsify it. These conjectures are all essentially another way of saying that the phenomena are not understood. This is a statement that rejects ideas which in the past have been imbued with much vigour and imagination, but it sadly remains the truth.

Another very feeble candidate for consideration lies in the modification of wave mechanics or of gravity which may be necessary for the two to be joined together. The chance of this marriage shaking the basis of wave mechanics so as to give macroscopic, or visible, effects which would explain the Geller phenomenon is so unlikely that we can dismiss it. So far no one has miraculously conceived of the kind of experiment necessary to give us any lead on how wave mechanics should be changed. A change of gravity is more likely to allow this unification. But even then it does not seem possible for a new gravitational theory to explain the Geller effect, since there would still be the seemingly insoluble problems raised in Chapter 6.

We have to return then to electromagnetism, our only hope, as the intentionality field. We must try to solve the problems facing it, especially those of energy transfer and focusing, and see how critical these difficulties really are. As a first step we must endeavour to pin down more precisely the possible range of wavelength or frequency which would be most suitable.

The intentionality field is able to penetrate into metal to at least a depth of a millimetre, if not more. This follows from the phenomenon of softening which has been directly observed a number of times during bending of metal – both specimens and spoons and forks. The process occurs very rapidly, in a few seconds or less, and the thickness of metal in these cases has been at least two millimetres, and sometimes slightly more. Instances, however, have been known of metal of considerably greater thickness becoming soft while bending. On the assumption that the intentionality field can penetrate the metal surface from all sides, the penetration depth of this field is therefore at least one millimetre, or it may be twice that if the intentionality field acts in one direction only.

We may use this penetration depth to put a severe limitation on the wavelength of electromagnetic radiation which might be causing the bending. The smaller this wavelength – and so the higher its frequency – the shorter the distance radiation can penetrate into the metal. This is because of the increasing amount of energy lost in the operation of causing the electrons in the metal to follow the undulations of the electromagnetic wave. The depth at which the electric intensity has become reduced by a half is called the skin depth; it is 1 centimetre in copper at 5 cycles a second but less than 1 millimetre in iron at the same frequency, because of its magnetic properties. There need not be a unique frequency involved here, but we can conclude from the results for material made of iron that frequencies no greater than 50 cycles must be involved, though in the case of copper they may also be as high as 150 cycles. These frequencies are well inside the range of those occurring in the body. The wavelength associated with such radiation is at least as great as 2,000 kilometres, and certainly includes a component with a wavelength three times that value. The difficult problem of focusing enough power with radiation of such an enormous wavelength on to a specimen has already been raised, as has the risk that someone else may pick up a telepathic signal. The answer to this problem made by Norbert Wiener, the founder of cybernetics, is appropriate here: to detect a signal carried by such waves would require an aerial ten thousand miles long!

It is not thought easy to detect any electromagnetic wave if the size of the object used in the detection process is much less than the wavelength of the radiation. And thus it would appear very difficult to achieve any very effective use in telepathy of electromagnetic radiation with a wavelength of thousands of kilometres. However, the emission of such radiation need not be the same in all directions. Aerials have directional properties, and the amount of energy emitted in different directions from a radiator can vary greatly with the direction. We can imagine that the curious directional effects discovered in the Geller effect arise from some feature in the human radiating system.

This hypothesis does not explain how power may be focused at certain points in the space around a subject and not at others. It also leaves unsolved the question of the source of the large amount of power needed to achieve the Geller effect. At this stage it is necessary to look in more detail into a mechanism which might produce distortion in a metal when exposed to low-frequency electric fields, notably how an incident electromagnetic wave of low frequency might cause the distortion. In particular we have to understand how the metallic bonds can be relaxed so as to produce the plasticity seen during several sessions of bending.

A metal, as stated earlier, is composed of a large number of small crystals called grains. Each crystal has a regular arrangement in which the charged metallic atoms form the points of a lattice through which the common electrons move almost freely. The metallic bond itself can be thought of as arising from the binding of an electron to the electrically attractive region encircling the oppositely charged metal atom. Such a structure has considerable strength, accompanied by ease of heat or electricity conduction allowed by the easy flow of the gas of electrons.

The metal bonds can be destroyed if the metal is heated to its melting temperature at several hundreds of degrees or higher. This heating can be accomplished by infra-red radiation, but since there is little temperature increase observed during the Geller effect this seems unlikely. It is conceivable that energy could be absorbed by the electrons in the metal at the natural frequency of their oscillations in the metal, called the plasma frequency, which is usually in the ultra-violet region of the electromagnetic spectrum. It would be difficult to prevent the lattice atoms from sharing in this energy gained by the electrons, so causing a temperature increase of the whole specimen. Even if this could be avoided we cannot substitute ultra-violet radiation as the intentionality field, as we have already presented direct evidence against its use in these circumstances. Another counter-argument is that such radiation would need to have a wavelength less than half that of blue light to excite the metallic electrons to their natural plasma frequency. But then it would not be able to travel very far in air – a millimetre or less. In addition the air would be broken down into charged particles on the passage of such radiation – it would be ionized. No ionization effect has been observed during metal-bending; the electroscope test described in Chapter 4 showed a negative result. It is further very unlikely that ultra-violet radiation would penetrate very far into a metal on account of the skin effect described earlier. In the case of copper, the skin depth for such ultra-violet radiation is only one-millionth of a millimetre, far too short to be of practical value in making a pronounced bend in a strip of metal a few millimetres thick.

For these reasons the Geller effect is unlikely to be occurring at an atomic level. Some consideration must therefore be given as to how bending could be achieved at a macroscopic level; often the fractures obtained by subjects look almost identical to those obtained by bending a piece of metal backwards and forwards until it breaks. In other cases it looks as if the metal was torn into two pieces, as was described earlier in Chapter 4. It is obviously necessary to consider afresh, but now in more detail, how conventional bending and breaking occurs in metals and similar polycrystalline materials.

The initial application of a force to a material causes a deformation, called a strain; when the force is removed the strain also disappears. Above a certain critical stress plastic deformation remains after the stress has been removed. This plastic deformation arises from the motion of imperfections in the crystal, the most important of which are dislocations. These were described earlier as lines marking the presence or absence of metal atoms, so causing a distortion of the regular lattice structure of the crystal. The movement of one of these lines produces a strain in the metal which permits some yielding to the applied force.

Dislocations are always present in a metal unless it has been very specially prepared. As we have seen, the number of dislocations can be reduced by heating the specimen to a temperature below its melting point. Bending a specimen by hand increases their number so that the stiffness of the material increases; the strength arises from the increasing intertwining of the dislocations and the reduced ease of their movements. It is possible to calculate how much energy a strip of metal may normally contain by the dislocations it possesses. Remarkably enough, one of the standard specimens could well possess enough energy to cause it to bend on its own. It does not, of course, do so. The problem is to discover how to produce movement of dislocations so that they actually do generate a visible distortion effect in the metal.

A possible mechanism is that of low-frequency oscillations being set up in the small crystals – the grains – of the metal. Electromagnetic radiation of a certain frequency will cause electrons to move following the varying electric field. If the boundaries of the grains present a slight resistance to electronic flow then this motion will bring a distorting force to bear on a grain and send it into slight oscillations. The resulting oscillations could then be transferred to the dislocations, and might cause them to move. Once enough dislocations were in close proximity to each other they would begin to coalesce, initially producing bending, then cracks, and finally fracture.

There may alternatively be a mechanism for direct movement of the dislocations. This idea also fits in that fractures tend to occur at regions with the greatest stress and, therefore, with the greatest number of already existing dislocations. The cases of fracture described in Chapter 4, in which no grain deformation was evident right up to the fracture surface can be explained on this basis, since the moving of dislocations may not alter grain size or shape.

It is necessary to work out in greater detail our proposed model of what happens, especially the movement of dislocations. One further important factor to be discovered is the amount of energy needed beyond that stored in the dislocations. This may be very low, as would be the corresponding electric field. It should be pointed out that this model may not be correct: it is only presented to indicate that the electromagnetic hypothesis of metal-bending is just viable as far as established science is concerned. The detailed mechanism we have described above is neither complete nor perhaps correct; only further experiments will show this. But that another related explanation may finally be found to be true is not out of the question.

It is clear that the electromagnetic hypothesis for metal-bending has a chance of being right, at least as far as the action of radiation on metal is concerned. We still have to consider the nature of the source of this radiation – the human operator. The same problem arises for telepathy and we shall try to discuss the two problems simultaneously, since both involve the sender in having some visual image in mind – that of a twisted piece of metal in the case of metal-bending, or the picture he is trying to transmit, in the case of telepathy. We should also expect the frequencies of electromagnetic radiation used for telepathy to be those generated by bodily processes, say between 1 and 5,000 cycles a second.

The wavelength of such radiation lies between 60 kilometres and 300,000 kilometres. The longer wavelength (lower frequency) radiation is particularly appealing since it would appear to have a dimension close to that of the earth (50,000 kilometres in circumference). Very long waves can travel right round the earth with little loss of energy. The surface of the earth and its atmosphere act together as a resonator when 'rung' by electromagnetic radiation from lightning storms. There is a sequence of best frequencies, the first three being 7.8, 14.1 and 20.3 cycles. These particular resonances were first observed in 1960 as the product of atmospheric storms. With these hardly any absorption occurs.

Nor is it only at special frequencies that long-wave propagation is most efficient. Radiation with a frequency of 5 cycles is reduced in intensity by only 5 per cent after travelling a distance of 10,000 kilometres. As the frequency increases the ability to travel great distances decreases. At a frequency of 75 cycles the reduction of intensity is 30 per cent; at 500 cycles 99 per cent reduction occurs at the end of 10,000 kilometres. Obviously the lower frequencies are more efficient for communicating over a long distance. It is these which are produced by the human body.

The crucial difficulty still to be faced is that of the reception of such signals. As remarked earlier, only short aerials are available in the human body and they afford a very low signal power. Detection of such long waves by a person would be very hard. In addition, there is very large interference at these frequencies from atmospheric storms. Very low power levels can, however, be detected by using suitable apparatus. Radio astronomers have been able to record the background radio noise from deep space down to 10^{-30} watts – a thousand billion billion billionth of a watt – by the use of large aerial systems. The amount of detectable energy being received in this way over a square centimetre is less than 10^{-40} watts. If a person is radiating a millionth of a watt at low frequency, say 5 cycles per second, then the power propagating through each square centimetre of area on the other side of the world would be about 10^{-24} watts. This might be detectable, but no radio astronomy techniques are available at these frequencies, the astronomical work involving infra-red radiation.

The reduction of intensity in low-frequency electric fields is very slight on propagation round the earth. We can conjecture that the brain or the muscle fibres might act as an aerial of sufficiently long length for such signals emitted by another person to be detected. It is possible to detect brain and muscular activity at a distance, though little work has been done on this. Directly transmitted waves may even activate nerve cells in other people. An aggregate of a thousand million nerve cells, each with an axon of a millimetre in length, could present a folded aerial with a total length of one thousand kilometres. There are enough of such nerve cells in the cerebral cortex – so the brain could possibly be sensitive to such long waves.

There are already indications that such low frequencies can activate the brain and body. Russian scientists have claimed that electromagnetic radiation of 50–100 cycles per second can stimulate isolated nerve and muscle tissue and also heighten the excitability of the central nervous system in animals and man. A conditioned reflex to radiation of 200 cycles has apparently been established in humans. These experiments will have to be repeated and extended to the lower frequencies of a few cycles per second. This applies particularly to radiation at the very low intensities emitted from the body.

Peripheral nerve cells respond to an external stimulus by changing their firing rate. We cannot yet be sure that this alteration of the firing frequency is the only method of communication for all nerve cells in the brain, though it is certainly known to occur in the primary sensory areas of the cerebral cortex. If we assume that this is how information is transmitted then we would expect electromagnetic radiation emitted by the body to have a variable frequency. Messages could be emitted if there were some degree of coherence in the firing activity of nerve cells in the cortex, as occurs during alpha or theta brain wave activity. So it is to be expected that telepathic transmission will occur most effectively in a mental state in which there is a clear-cut EEG pattern.

There is evidence to support this. Stanley Krippner and his associates at the Maimonides Dream Research Laboratory in New York have had success in detecting variation of telepathic powers according to the state of consciousness. Both during use of a machine that induces sleep artificially and also in periods of dreaming, it has been possible sometimes to transmit quite clear visual images. In one instance a group of two thousand senders attending rock-and-roll concerts were asked to concentrate on 'sending' randomly selected art prints to a particular subject asleep at the Maimonides Dream Laboratory. A second subject's dreams were also monitored; only the named subject had dreams significantly related to the prints looked at by the senders. When the sleep machine was used the results were just as successful. This machine sent people off to sleep by administering small electrical pulses at 100 cycles per second to subjects through electrodes attached behind the ears and to the eyelids. When the subject was put to sleep in this way, he tended more often to be right about the contents of a sealed envelope than when he was awake and distracted by being exposed, say, to flickering lights.

It is known that not only can the electrical activity of the surface cells of the brain be altered by training, but also that of deeper-lying tissue; dogs have been trained so as to be able to alter the 3–6 cycle theta brain waves found in regions below the cerebral cortex. So when it comes to the process of direct electromagnetic radiation of information we cannot exclude the deeper-lying units from involvement. This widens the field and admits of greater possibilities.

In short, the electromagnetic hypothesis for the intentionality field occurring in the Geller phenomenon still appears to be feasible. The mechanism involving movement of dislocations suggested as an explanation of metal-bending does not require an exorbitant amount of energy. Nor does the emission, transmission and reception of electromagnetic radiation by the brain appear completely impossible, at least with our present understanding of its nervous activity. What holds us up in either proving or disproving the electromagnetic theory is our ignorance of many of the processes involved in the Geller phenomenon. More experiments are needed to test the theory.

First of all, the conjectured mechanism for metal-bending must be verified. Suitable instruments attached to specimens could indicate any oscillations produced in the metal itself. The distribution of dislocations in specimens at different stages of bending is also extremely relevant. Finally, it is necessary to show that such a process can be fully simulated under laboratory conditions by a suitably constructed electro-mechanical 'stroker': only when it is possible actually to duplicate the distorting of metal shall we know how the Geller effect works.

As far as the electromagnetic hypothesis itself goes, highly sensitive detectors for monitoring the low-frequency emission from subjects, both during metal-bending and telepathy and at other times, will be required. This field of 'remote electroencephalograph' analysis is as yet completely unexplored. It will surely contribute towards the solving of the particular problems discussed in this book, and also to our general understanding of the human organism.

9 The implications for society

There is much more to ESP than bending cutlery or sending simple visual images from one person to another. The powers and skills of mediums and astrologers, the behaviour of ghosts and poltergeists, out-of-the-body experiences, voices of the dead picked up on tape recordings, precognition and materialization of objects – these are all major aspects so far not discussed. The reason for keeping strictly within the limits of the Geller phenomenon has, I hope, become clear in the course of the preceding chapters. The manifestation of Geller's powers has meant that for the first time a serious attempt now seems possible to find a scientific explanation for one distinct extrasensory phenomenon. We can then move on to see whether any of the mechanisms thought to be at work in the Geller phenomenon could have any bearing on all the other phenomena.

Some of these other forms of ESP could never be explicable in terms of present-day science. Precognition is one of these, involving as it does the violation of causality by the sending of messages into the past. Nor is the materialization of objects, such as camera cases or filmy clothlike substances (a favourite with many mediums of the past), to be reconciled with modern science; the amount of energy required for such feats is too large. But when we turn to poltergeists, ghosts and out-of-the-body experiences a glimmer of scientific credibility emerges. This is not to say that all such instances could be scientifically explained, but at least we may soon be able to say that some are *bona fide* cases.

The phenomenon of poltergeists was famous in antiquity, and one instance occurred as early as AD 856. In general, the poltergeist has proved noisy, manifesting itself as the sound of bells being rung, footsteps or other noises. A classic case is the Pealing Bells of 1834, which involved the mysterious ringing of nine old household bells. Those bells could be actually seen ringing, but other cases have been of more puzzling sounds. The phenomenon may also involve the movement of objects, sometimes with great force, and sometimes even making holes in walls or ceilings.

A typical account of a poltergeist phenomenon is given by an associate of the Society for Psychical Research: 'On this Sunday evening occurred the really most unaccountable noise of any yet noticed. I was in the house alone, writing at my desk. Time 3.30. Suddenly I heard a noise which seemed to come from the hall, outside my room door. I can only compare the sound to that which would be made if half a brick were tied to a piece of string and jerked about over the linoleum – as one might jerk a reel to make a kitten playful.'

A common feature of poltergeist phenomena is that they are usually associated with children, especially young girls. It is possible to understand the occurrences in terms of the electromagnetic hypothesis; children appear to have far more powerful electromagnetic intentionality fields than adults. The explanation then is along the lines of that offered for the movement of objects achieved by Kulagina or Vinogradova. This was not considered in the previous chapter, but results of experiments on these two subjects by Russian scientists, described in Chapter 3, indicate that some form of electromagnetic field may be involved. In Vinogradova's case strong electrostatic fields were observed during experiments, while Kulagina had an associated low-frequency magnetic field. These fields may produce motion in light objects by a separation of positive and negative charges on the objects, in a way which may be similar to that in the Geller effect. It is still difficult to understand how enough energy can be radiated from a human brain to cause, for instance, an object to be hurled across a room. On these grounds alone one is tempted to doubt the truth of such extreme cases and to accept only those instances where gradual movement of objects with much lesser amounts of energy and without physical intervention may actually have occurred.

Ghosts have had their role in human affairs almost from the beginning of recorded time, as their frequent appearance in fiction or in drama illustrates. An extensive study of reported apparitions was made at the end of the last century under the auspices of the Society for Psychical Research. The phantasms were classified in two groups: those resembling people still alive who were undergoing a crisis at the time (more often than not, they were dying), or those resembling people who had been dead for more than twelve hours.

Seven hundred and two cases in the former category were analysed, and all seemed to be interpretable in terms of receipt of a telepathic message from the dying agent by someone especially close to them. This explanation is supported by the fact that these ghosts have a way of vanishing without trace, and occasionally behaving in ways impossible for physical beings: thus they may instantaneously disappear inside a locked room, or even walk through walls.

A typical case occurred in 1884 and was recorded as follows by the person experiencing it: 'I sat one evening reading, when looking up from my book I distinctly saw a school-friend of mine, to whom I was very much attached, standing near the door. I was about to exclaim at the strangeness of her visit, when, to my horror, there was no sight of anyone in the room but my mother. I related what I had seen to her, knowing she could not have seen as she was sitting with her back towards the door, nor did she hear anything unusual, and was greatly amused at my scare, suggesting I had read too much or been dreaming.

'A day or so after the strange event I had news to say my friend was no more. The strange part was that I did not even know she was ill, much less in danger, so could not have felt anxious at the time on her account, but may have been thinking of her; that I cannot testify. Her illness was short, and death very unexpected. Her mother told me she spoke of me not long before she died . . . She died the same evening, and about the same time that I saw her vision.' The lady who had this experience added that it was the only hallucination she had ever had.

Similar cases went on occurring up to the present day. The careful analysis of the 702 cases by the Society for Psychical Research indicated that very few of them could be explained away as aberrations. We can conjecture that these phantasms of the living are explicable in terms of the electromagnetic aura of the dying or otherwise stressed person becoming stronger at the time of the crisis, and so exciting an hallucinatory response in those especially responsive to that particular radiation. As with telepathy we have no clear indication of the details of how such information is transmitted. But this is no reason for thinking they cannot be discovered by further research.

Clearly a different problem arises with the other class of ghost, the phantasm of the dead. Some 370 cases of this kind were investigated by the Society for Psychical Research at the end of the last century. Most of the instances were almost the opposite of the blood-curdling apparitions of folklore. The real-life ghost is usually lifelike in appearance and is to be seen around a building, sometimes by someone apparently innocent of previous ghostly appearances. Occasionally the vision appears to more than one person at once.

A description of one such apparition, seen in the same situation but at a different time by the previous reporter's daughter, reads as follows: 'I was suddenly aware of a figure peeping round the corner of the folding-doors to my left; thinking it must be a visitor, I jumped up and went into the passage, but no one was there, and the hall door, which was half glass, was shut. I only saw the upper half of the figure, which was that of a tall man, with a very pale face and dark hair and moustache. The impression lasted only a second or two, but I saw the face so distinctly that to this day I should recognise it if I met it in a crowd. It had a sorrowful expression. It was impossible for anyone to come into the house without being seen or heard.'

One of the investigators into these phantasms of the dead considered that a good proportion could best be understood as 'manifestations of persistent personal energy', by which he meant only that the apparitions had some connection with a person who had once been alive. If so many instances are to be taken seriously, they could be explicable along the lines of the Geller effect: by a person causing a distortion of the particular environment in which he experienced great emotional stress. His electromagnetic aura could have left an indelible mark on parts of the physical surroundings – the walls, ceilings, floor and so on – of where he had been at the time. Another person could cause his own electromagnetic field to be reflected back by this distortion in his surroundings so as to produce the visible apparition, rather as a distorting mirror at a fair can produce several different images of one person.

This hypothesis would explain why more than one person could see a ghost at the same time, since numbers of people may be simultaneously influenced in this way. It can also explain why the ghost would not be visible to certain other people – especially sceptics – because their electromagnetic auras would be too weak or otherwise not suitable to be reflected by the environment and so produce the hallucination. Nor would a camera ever be able to photograph a ghost, of the living or the dead, since both kinds of image exist only in the mind of the beholder.

Out-of-the-body experiences have been claimed for many years. Some of these are extremely vivid, and either involve an association with an alternative body or are experiences with no spatial extension at all. Thus one person remarked, 'I am disembodied, but in a small space which has a definite size and location'; or another: 'I was like a sheet of paper floating above my body'; yet another, even more apposite to our interpretation: 'It was not another body; perhaps something more like a magnetic or electric field.'

There are cases where information has been obtained during an out-of-the-body experience which could not have been obtained by any normal means. One subject writes: '. . . I looked down at my body, then kind of floated out of the room into some street I didn't know and stopped before a house and I entered and went to a bedroom facing the stairs. The man lying in bed was a very old friend whom I had not seen for a year or two. I met him about six or seven months later and he said he had moved to a new district. When I told him he lived in an upstairs flat and how the furniture was placed in the bedroom he wanted to know how I knew . . .'

These and similar cases, some personally recounted to me, have an air of authenticity. Experiments are now in process which may produce really firm supporting evidence. If they succeed the only mechanism likely to be operating is the electromagnetic aura. The idea of extending one's normal vision by means of electromagnetic waves *seems* unlikely, but *may* just be possible. Further experiments with sensitive detectors of electromagnetic radiation need to be carried out on subjects as they are having an out-of-the-body experience, and should, with other tests, help to give definition to the phenomenon.

When we turn to mediums, we have a whole range of paranormal phenomena – fortune-telling, messages from the dead, the moving of objects, materializations, information recounted about the sitters that is normally unavailable to the medium, and so on. The truth about precognition has already been declared very difficult to establish by the scientific method, so that the successes of the fortune-teller may either be purely at the level of chance, or be related to the knowledge obtainable from a sitter without his or her knowledge. This comes under the heading of telepathy, though in a very acute form; it might still just occur by means of electromagnetic aura.

The materialization of objects by a medium has also been remarked on as being scientifically very difficult to prove, while any success in contacting the dead could again be explained by the 'mind-reading' ability of the medium. I was once seriously advised before visiting a medium in order to test her powers: 'Guard your mind.' But then in so doing I may have prevented her telepathic faculty from performing at its most effective level.

Not long ago the voices caught by tape recorders in what would otherwise be silent rooms caused a stir in the scientific establishment. Seventy thousand conversations obtained by this and similar means by the Latvian psychologist Konstantin Raudive have been found to have certain strange properties. In particular there is apparently a 'reply' to questions put by Raudive and his collaborators. There are certainly sounds on tape, but they are too indistinct for any one interpretation to be definitive. Raudive himself claims that man 'carries within himself the ability to contact his friends on earth when he has passed through the transition of death.' But the phenomenon itself is more simply and rationally explained as radio waves picked up from a wide variety of transmitters, as well as being perhaps effects from the electromagnetic aura of others. Tests with carefully designed shielding should be able to show whether or not electromagnetism has any real role in this as the mechanism.

Astrology since time immemorial has enjoyed great popular appeal. The argument is that the influence of the planets and the stars is of great significance. But the effects of any radiant energy from the planets or stars other than our own sun or moon are quantitatively very slight indeed. Jupiter, for example, exerts a gravitational attraction of less than one ten-thousandth that of the sun. What are of much greater importance in human development are atmospheric storms, temperature variations and other physical influences described in Chapter 7. These could exert their own characteristic modifications on genetic and environmental personality traits. This might go some way to unravel how much success astrology has in correlating character with the twelve signs of the Zodiac.

Evidence has been presented in the first four chapters which shows that people exist with the powers of bending metal in an inexplicable manner, and of telepathy: these are the 'superminds' of the title. The performances of these superminds has been observed often and under conditions which exclude fraud. Some of the detailed effects seen in the bent metal could not in any case have been achieved by deception, but only by means of some hitherto unknown process. The telepathic feats also come under this heading: information is detected by a supermind in a manner which baffles the understanding.

All this appears to postulate the workings of a so-called psychic power and psychical interpretations have been cogently advanced by investigators who put great emphasis on the non-physical aspects in their explanations. This approach was popular with many of the earlier workers in the field of ESP, yet it automatically excludes any step towards physical analysis and models for the processes. If the powers of the superminds in our midst are to be understood, we must take the scientific path, and trust that the physical world is sufficiently elaborate to furnish us with an explanation. If we do not try to come to terms with the abilities of the superminds, then cults of unreason will flourish from the seeds sown.

Already Geller has been hailed as the lone human contact with a galactic civilization of advanced powers. But the spoken messages that have allegedly been received have been banal, and at times totally inaccurate in relation to what we know. One's sympathies lie with that statement of T. H. Huxley: 'Better live a crossing sweeper than die and be made to talk twaddle by a "medium" hired at a guinea a séance.' The other evidence brought forward for such contact is that of the bending of cutlery or the mending of watches, and the dematerialization and materialization of various objects belonging to investigators. The first two of these phenomena could be regarded as beneath the powers of emissaries of the galactic civilization, if one exists. But these effects are now being duplicated by children in various countries throughout the world, *without* evidence of galactic contact. The third process – transportation of objects and, it is claimed, even of human beings, especially in and out of completely enclosed spaces – is impossible to fit into modern science. Unfortunately this is the kind of thing that many people want to believe in, even without proper evidence. As modern science has performed the so-called 'miracles' of lunar probes and heart transplants, is it not odd to find so many of us still unsatisfied and clamouring for a convincing exhibition of supernatural powers?

Yet in some sense the extra-terrestrial explanation of the Geller phenomenon is grounded on a kind of reasoning. There may well be civilizations which have developed on distant planets millions of years before ours did. By now their science and technology could be such as to allow for extensive space exploration. They may at present be attempting to make contact with us by what appear to us to be strange methods but ones which may have been chosen for reasons we cannot yet comprehend: not very likely, of course, but just possible.

Many other ideas have been advanced for the Geller phenomenon. Some of these alternative candidates are of a nature to allow for more careful assessment, and just such an analysis has been made in the four preceding chapters. Only one mechanism likely to be mediating either metal-bending or telepathy remains out of the host of scientifically based models for the intentionality field – this is electromagnetism. We have seen, in the previous section, how the suggested mechanisms of the intentionality field in conjunction with metal-bending can help to explain a range of other extrasensory phenomena. This gives extra support to our conjectures but, of course, it does not prove them. This has yet to be done. Let us suppose we are correct in our assumptions. It will then considerably reduce support for the extra-terrestrial communication hypothesis, except in the case of the transportation of objects. This has, in any case, proved in no state to be analysed scientifically, being highly non-repeatable. Science can do little with such events until they can be controlled.

Several interesting conclusions can be drawn if the hypotheses of the previous chapter are correct, of which the foremost is that the present scientific model of the universe can be used to explain a hitherto far-removed area of human experience. This scientific underpinning of so much of ESP may even help to reconcile many men and women to the apparent divorce of modern science from their common experience. Strange phenomena have been experienced by a fair proportion of the population, but in the past science has categorically denied their validity. This has only served to bring the scientific method into disrepute in the eyes of many. It is my hope that science will in fact become more all-embracing.

At the same time, however, the understanding gained has its threatening aspect, in that by explaining away mediumship and ghosts in our present fashion, we are going a long way towards destroying the possibility of life after death. The only scrap of non-religious evidence for survival after death is contained in messages from mediums or on tape recorders. This evidence needs to be proof against explanation by telepathy via the medium, or by electromagnetic radiation acting on the recording machine, and should only concern events which no one (except the dead) could have known about. There is very little, if any, information transmitted which satisfies these criteria. This leaves little for those who were hoping for a better life hereafter, and is cold comfort for those close to many nearing the end of their lives. If the mechanisms we conjectured to be at work are proved correct, it will be necessary to present these particular implications as widely as possible, so that when death comes its nature will be clearly appreciated. If there is life after death it can only be by absorption in God in a manner passing all comprehension; this would accord most closely with modern theological thinking. But even if there is anything on the other side, death is still a barrier.

If ESP is explicable by electromagnetic radiation, we may have to alter our ideas on the nature of mind. It is often said that the mind is more than a set of active nerve cells and their connections. We see that this is indeed the case on the basis of our present hypothesis, in which the mind is to be regarded as extending in space as far as its electromagnetic aura. It is not clear if the aura can have any effective reaction back to the physical brain producing it, but there is no obvious reason why it could not. The mind should thus be regarded as consisting of electromagnetic activity both in the external aura and in the internal network of nerve cells; the brain, regarded as a structure of relatively static molecules, forms this receiving net. The title of 'superminds' would then be apt indeed!

This model of the mind is a physical one, and we see here another important result of the electromagnetic hypothesis concerning the Geller phenomenon. It has led to a materialistic interpretation of the universe, both of matter and mind. The crucial problem of explaining how we obtain our private mental world from a set of purely physical quantities is not an easy one, though a tentative solution in terms of a comparison theory of mind, where thought has its detailed content derived from a comparison of stored relevant knowledge, has been given elsewhere.* If this formulation can be made, the materialists' hypothesis is again proved correct and the universe shown to be composed of forms of energy.

Then what about religion, we may ask? It still has a place in our lives, as it does not depend for its validity on the detailed nature of the physical world. Its appeal lies more in dealing with the difficulties raised by the question: 'What is the reason for the existence of the universe?' To this science can never give an answer; only through the concept of a completely unknowable God can we come to terms with it.

To return to more immediate possibilities, a rediscovered form of communication between people, outside the usual five senses, and a different method of working metals, these hold promise of many different applications, some of them good and some bad – in the moral sense. Let us suppose that the electromagnetic aura became a reality. Its exploitation might be found to be reasonably

*J. G. Taylor: *The Shape of Minds to Come.*

simple, for example by means of a powerful enough aerial system. A whole range of visual images could then be impressed on an unsuspecting populace, especially that of another power. Clearly, once such a process is scientifically understood, its technological development is merely a matter of the resources made available to suitable scientific teams. Already the powers of superminds are being used in many practical situations. Dowsers are regularly employed by large corporations for finding water, oil and mineral resources. Clairvoyants are used by the police to help find missing people. The Dutch clairvoyant Gerard Croiset has had some astounding successes in discovering the whereabouts of graves of soldiers killed in action in the last world war. Official interest in the powers of superminds seems to be growing – the American Federal government is reported to be spending about a million dollars a year on research in this field; it is also providing over fifteen million dollars a year on research into how microwaves and radio waves affect living things.

And what about the effect on computers when superminds get working on them, the way that Uri Geller and Ingo Swann have managed to? Will crucial links in the defence chains of various governments be at risk? There may well be grounds for the view of a group of defence workers, so the apocryphal story goes, who advised their government to liquidate all superminds as being a menace to security. Clearly they have met their match in the eight-year-old girl who can wipe magnetic tapes clean from a distance. How useful she could have been over Watergate! The Russians, who allocate millions of roubles a year to ESP research, may have had the dangers of misuse of such power in mind when they sentenced E. K. Naumov, the doyen of Russian parapsychology, to two years' hard labour on a clearly trumped-up charge. Naumov had had continual contact with Westerners – could the Russians perhaps have made a breakthrough which could only be kept secret by putting Naumov out of circulation?

But the powers of the superminds in our midst can be of great value, if used wisely. The scientific understanding of these abilities could ultimately lead to amplified emission of radiation carrying information only detectable by a few. In the same way a new range of metallurgical processes could be developed which would allow cold-working of metal using far less power than is needed for present-day techniques. The possibilities here are enormous, including the chance that the non-superminded might be able to develop their powers using suitable biofeedback monitors.

One aspect of this could be the effectiveness of communal mental effort, such as one sometimes sees in religious congregations. Gatherings of this kind could bring benefit to the participants by the interaction of their individual electromagnetic auras. Large gatherings, in particular, might increase the total radiation so that 'miracles' of healing could take place through the interaction of radiation with living material. On the reverse side of the coin, gatherings could produce extremes of anti-social behaviour, like those at mass demonstrations and football matches. How useful it would be if low-frequency radiation could be employed to reduce rather than enhance certain unpleasant behaviour patterns! This way lies an interesting weapon for the future.

At the end of the book we are only at our beginning. The actual experiments must now be carried out to verify the conjectures. With luck we will discover that missing physical mechanism. And more likely than not it will be found on earth, rather than in the heavenly galaxy.

10 Postscript

It is now eight months since I first saw Uri Geller demonstrating the 'impossible'. This record is my attempt to make sense of what I saw. I have spared my readers details of long hours spent visiting new subjects in their homes up and down the country, and observing them while they tried to demonstrate powers, or failed to do so when prying cameras or scientific measuring instruments proved too much for their self-confidence.

I did go on, however, for many months in the hope that with practice subjects would improve their metal-bending powers. To a certain extent this did happen. But I was not able to convince my fellow scientists. One distinguished scientist, a Nobel Prize-winner, told me that metal-bending was clearly done by fraud, and his wife threw in for good measure that no scientist of repute would be caught dead investigating Geller. A scientific colleague with great research funds at his disposal would not hear of the effect being possible. In any case, investigation could not be supported financially; 'Suppose a question were asked about it in the House!'

The refusal to take Geller seriously is not confined to scientists. The mass media are full of stories about Geller being exposed or of chemicals or conjurers able to achieve the same effects. I hope that the evidence presented in this book will go some way towards convincing the general public that the Geller effect is genuine and that we *need* to understand it. I hold no great hopes of denting the scientific complacency of my brethren. Only a machine able to perform equal miracles could do that. But a machine needs a blueprint before it can be built, and without knowledge of the mechanism by which metal-bending occurs no such blueprint can be devised.

We have in fact lately tried to build a mechanical metal-bender. This consisted of a pad, attached to a motor-driven piston, which was made to rub back and forth on a metal strip. Both the speed of rubbing and the pressure used could be controlled; the method of applying pressure used was an exact imitation of that used by several metal-benders. The purpose of the machine was to see if the actual method of rubbing was responsible for setting up a special type of vibration in the metal, so bending it. After many hours of operation, with speeds (around 2 cycles per second) and pressure (about 20 grammes) typical of subjects, no deformation was observed, beyond 1° of bend on the standard specimens.

Further clues about the Geller effect were clearly essential if we were to avoid building one unsuccessful machine after another. Perhaps, I thought, we might pursue the idea of the electromagnetic mechanism presented in so favourable a light in our earlier conclusions. But the unresolved questions – How could electromagnetism react with a metal to cause such distortion? And how could it be emitted and focused at will by a human being? – still remained stumbling blocks. At any rate, the tests devised specifically to show the crucial influence of electromagnetism proved unsuccessful. So the search has gone on for new subjects and for new types of tests to try the well-established metal-benders, in the hope that sooner or later a useful clue would turn up. And this is what happened; although the situation became worse before it got better.

Two unexpected events were responsible for the change. The first of these was a three-hour visit to my laboratory by Uri Geller on 20 June 1974. It produced results which enormously widened the range of phenomena, and gave a very clear validation of Geller's ability to distort a wide range of materials. The second was the appearance of two new subjects, both boys and aged ten and sixteen years, who had very strong powers as metal-benders, the elder one even rating as highly as Geller himself. The sixteen-year-old was, for example, able to bend a straight strip of aluminium, about 18 centimetres long and sealed securely in a Perspex tube about 2 centimetres in diameter, into an S-shape: quite impossible, it would appear, without taking the strip out of the tube and then bending it mechanically. Yet careful scrutiny showed that the seals on the tube had not been tampered with. This further proof of the Geller effect helped to convince some of my scientific colleagues, especially because these 'miraculous' results were obtained on a number of occasions. But how it was done remained a mystery, and only when tests were made for other powers, in particular moving objects and not just bending them, was new light thrown on the Geller effect. At the same time it was revealed that the human control over materials was much more powerful than had been expected; even the possibility that objects could be removed from sealed boxes had to be seriously considered.

Geller only gave me twenty-four hours' notice, but since I already had various experiments prepared for other subjects this caused little bother. In an office at King's College I had set up several experiments designed to measure the pressure applied by Geller during metal-bending. The two principal ones were very simple. The essential apparatus for one of them was a balance of the type used to weigh letters and parcels, sensitive enough to measure weights to a quarter of an ounce. A brass strip about 20 centimetres long was taped horizontally to the platform of the balance. The major portion of the strip extended out from the platform, and Geller stroked the top surface of it while I measured, both directly by reading the scale, and using an automatic recording device, the pressure he was applying. At the end of the test the strip had acquired a bend of 10° although Geller had at no time applied more than half an ounce (20 grammes) of pressure. It was out of the question that such a small pressure could have produced that deflection. What is more, the actual bending occurred upwards – *against* the pressure of the finger. Earlier another subject gave a similar result, producing a smaller upward deflection (2°) on a strip of copper with less than an ounce of downward pressure.

While Geller was doing this experiment, it was a little disconcerting, to say the least, to have the needle indicating the amount of pressure on the letter balance, also bending as it did through 70°. This didn't seem to upset the operation of the balance, though it did make the reading of the scale a little difficult. But the more devastating was yet to come.

The apparatus for the other test was a small cylinder embedded in a strip of aluminium in such a way that one end of the cylinder, covered by a pressure-sensitive diaphragm, was flush with the surface of the strip. When pressure was applied to the diaphragm in rubbing the strip gently with a finger, an electric current of amount proportional to this pressure was generated by a device installed inside the cylinder. This pressure-measuring device had been used with various subjects, but no bending had been achieved. In Geller's case the consequences were drastic. While holding the strip in one hand he made it bend in the appropriate region so that the pressure could be measured. But as the bending occurred the mechanism in the cylinder suddenly stopped functioning. I took the apparatus from Geller and observed, to my horror, the pressure-sensitive diaphragm begin to crumble. A small hole appeared in its centre and spread across its whole surface till the diaphragm had completely disintegrated; the entire process only took about ten seconds. After another three minutes the strip in which the cylinder was embedded had bent a further 30°. The Geller effect had been validated, but at the cost of £200 worth of equipment!

Attempts to influence objects without contact yielded more information. Geller held his hands over a plastic container in which had been placed a small crystal of lithium fluoride; within ten seconds the crystal broke into a number of pieces. There was absolutely no chance of Geller having touched the crystal: throughout the experiment I could see a gap between his hands and the container holding the crystal. He also buckled a small disc of aluminium, again inside a plastic container, while I held my hands between Geller's and the container in order to prevent any possibility of his directly manipulating the disc.

Geller was then led into another room to work with other pieces of apparatus. One of these was a strip of copper on which was glued a very thin wire. Distortion of the strip would cause a change in the electrical properties of the thin wire, which could be measured very accurately. Geller tried to bend the copper strip without direct contact, but had not done so after several minutes and there was no significant change in the properties of the thin wire. We broke off in order to start measuring his electrical output, but turning round a few moments later I saw that the strip had been bent and the thin wire was broken.

Almost simultaneously I noticed that a strip of brass on the other side of the laboratory had also become bent. I had placed that strip there a few minutes before, making sure at that time that it was quite straight. I pointed out to Geller what had happened, only to hear a metallic crash from the far end of the laboratory, twenty feet away. There, on the floor by the far door, was the bent piece of brass. Again I turned back, whereupon there was another crash. A small piece of copper which had earlier been lying near the bent brass strip on the table had followed its companion to the far door. Before I knew what had happened I was struck on the back of the legs by a Perspex tube in which had been sealed an iron rod. The tube had also been lying on the table. It was now lying at my feet with the rod bent as much as the container would allow.

Pandora's box had certainly opened up! None of the flying objects could have actually been thrown by Geller as he was some distance away from them and would not have been able to get close to them without being spotted. I was not wholly surprised as an earlier occurrence in the corridor had led me to expect some phenomenon of the sort might happen; I had been walking along a corridor with Geller from my office after the first series of tests when a strip of metal which had been left on the desk in my office suddenly fell at my feet. We were at least seventy feet from the office. I have to admit Geller *might* have brought the thing out of the room with him, but anyway I had been alerted.

To check on this under more repeatable conditions, I set a compass on a stable surface and asked Geller to try to cause the needle to rotate without touching it. This he did by passing his hands over it, achieving a rotation of up to 40°. Then I tried to do the same, keeping 10 centimetres away from the compass as Geller had. It proved impossible, either by imitating his movements or by stamping on the ground. Even rocking or rotating the compass directly had little effect except when obvious effort was used. Nor could Geller have been using a magnet unless he could palm it with consummate skill at particular moments, for he appeared to be able to switch on and off his effect on the magnet at will, in spite of the fact that he was making similar hand movements. Nor could my two companion observers detect any such deceit.

The next step was to make further tests, especially to see if non-magnetic material could be moved. But unfortunately Geller's time table didn't allow this. Right at the end of the session a comparatively loud click was heard at the far end of the laboratory. Looking towards its source we discovered that the small piece of metal which had flown to the far end of the laboratory was no longer lying on the floor. We searched the laboratory, but it was nowhere in sight. Geller remarked that this was not the first time things around him had disappeared; the piece of metal had most likely vanished from the laboratory. After he left I made a more thorough search of the room, and finally found the piece of metal under a radiator at the opposite end of the room from where it had been before. How it had got there I do not know, but it clearly had not dematerialized as Geller had suggested.

This left me in a state of even greater mystification than before. The bending of metal by unknown means had been shown to occur, as had the distortion of other materials. But objects had apparently been made to 'fly' through the air, and a compass needle had been caused to rotate without the intervention of a visible mechanism. These events seemed impossible to comprehend; I should certainly have dismissed reports of them as nonsense if I had not seen them happen for myself. I could always take the safe line that Geller *must* have been cheating, possibly by putting me into a trance. I had no video-tape to support my own direct observations, though other people had seen the rotation of the compass needle. Yet I was perfectly well able at the time to monitor various pieces of scientific equipment while these objects were 'in flight'. I certainly did not feel as if I was in an altered state of consciousness.

Some understanding of these new phenomena could be achieved if there were a link between the moving of objects and the bending of them. If one thinks of it, what would be easier for someone who can cause cutlery and other objects to bend and break than to move them, even up into the air? These new poltergeist-like phenomena might not be so strange after all – or at least no stranger than metal-bending; provided, of course, that objects did not disappear at will.

Geller had enabled me to see that the repeatable physical phenomena had more to them than the metal-bending aspect. Just how much more would depend on observing the phenomena more meticulously. Here the arrival of two new subjects allowed new factors to be discovered.

In the middle of June I had a letter from the father of a ten-year-old about his son's powers, revealing a new departure in metal-bending known as the 'scrunch'. This was achieved by placing a number of straightened paper-clips in a cardboard box and getting the subject to think of them folding over and over on themselves, or 'scrunching'. I tried this with the subject and a sixteen-year-old friend, sealing a cardboard box with tape. Within half an hour several of the paper-clips were folded over and over in the typical scrunch formation. A further test used three straightened paper-clips, two of which had had their internal stresses removed (by annealing) to different extents. The resulting bending of each of the clips was through an amount which was largest for the clip which had been untreated and so had the largest amount of internal stress; least bending occurred for the one which had had most of its internal stresses removed. Here was a very valuable clue – bending depended on the existence of internal stresses in material. It agreed with the observation that in a spoon or fork the bending nearly always occurred at the region of greatest stress, at the handle. Precisely how the internal stresses were used was not at all clear.

Progress came in understanding this with the test involving the deflection of the compass needle. When tried with the sixteen-year-old it worked beautifully. Time and time again, the needle moved round and round. Not that this test was always successful, but it was more dependable than metal-bending. Was this to be our repeatable phenomenon – the answer to the psychic researcher's dream?

Many questions could be asked about this phenomenon. But first of all, it was necessary to devise a suitable suspension system, so that various types of material could be rotated at will with a minimum of external force required to produce observable motion. The simplest arrangement was to suspend a rod by a thread in a transparent cylinder. This cylinder was constructed with close-fitting ends to prevent any effects from draughts. Several cylinders were made and used successfully, thin rods of iron, copper, brass and Perspex all being caused to move through varying amounts when suspended in the cylinder. Fraud was clearly eliminated by the moving of non-magnetic materials, since no direct contact was possible and palmed magnets would have been powerless.

For iron rods, the amount of rotation achieved appeared to depend critically on the amount of stress present in the material; no rotation was achieved with rods which had all stress removed. Exactly the same dependence on stress was found with the straightened paper-clips (containing varying levels of stress), as in the case of bending described earlier; the greater the amount of stress the greater the rotation possible. The essential role played by stress in both rotation and bending gives a hint that these two phenomena may involve the same mechanism.

There is still the very difficult question as to the nature of the intentionality field which achieves this bending or rotation. In the case of the rotation experiments using the cylinders, some reasonably clear indications have come to light. No magnetic field was detected down to a level of a thousandth of that of the earth's magnetic field while rotating was occurring. However, the presence of an electric field during rotation was verified both directly, using sensitive receivers, and indirectly. One test involved finding out how rapidly the amount of rotation which could be caused decreased as the subjects moved further from the cylinder. Another test consisted of inserting sheets of metal between the arm of the subject and the material being rotated, to see whether increasing the thickness of such shielding reduced the amount of rotation. The results of both of these tests showed conclusively that low-frequency electric fields were involved, in that the frequency range agreed very well with that obtained by direct measurement. That the effect was not magnetic was also supported by the fact that applied magnetic fields caused amounts of rotation of the iron rods, which did not change much on loss of stress; no rotation could be achieved of non-magnetic materials.

One feature of the rotation experiments which allowed these indirect tests to be made was the highly localized emission of the intentionality field from the subject's arm. The apparently active area was a disc no more than two or three centimetres across about the same distance above the wrist on the inner arm: significantly, this same arm was the most effective one in metal-bending. At times the subject would complain that sensations of static electricity were building up over this arm; to relieve these he would put his hand in water, sometimes experiencing an electric shock when he did. All of this clearly supports the electrical nature of the phenomena.

Curiously, the eczema suffered by the subject itself gave some insight into the origin of the intentionality field causing metal-bending. This was particularly severe on his arm during the rotation session and especially over the apparently active region on this forearm. The amount of irritation caused by the eczema appeared to be closely correlated with the power of rotation the subject had over the suspended rod; when his arm had soothing ointment on it his powers were absent. One might speculate that this irritation was caused by the emitted radiation interacting with the surface of the skin. The same irritation was often experienced by the subject when he was bending metal objects, but on those occasions the sensitive areas were located on the upper part of the spine, and behind the knees as well as on the arm. This leads one to suppose that the electromagnetic radiation causing bending was being emitted by various muscles in the body and that the relative positions and movement of different parts of the body perhaps determined the form of the distorting field.

The hypothesis is that metal-bending and rotation are both caused by low-frequency electric fields emitted by various parts of the body, these fields being amplified by stresses in the materials themselves. Evidence has already been presented in support of this hypothesis. There is also some evidence from work on electrical breakdown in crystals which indicates that stresses in materials can amplify electric fields. What is clear is that much careful experimental and theoretical work will have to be done to test further and to amplify the hypothesis; research is needed in designing machines which can achieve the same effects. These findings give hope that such endeavours may ultimately be successful.

In the last chapter the electromagnetic explanation of metal-bending was extended to various other psychic phenomena, for example, poltergeists, ghosts, out-of-the-body experiences, mediumship. One field of great current interest is psychic healing, and it is appropriate here to consider the possible relationship between it and the Geller effect. If there is a link, then the electromagnetic explanation of the Geller effect should apply to psychic healing. The enormous value to mankind of such a discovery is obvious. New techniques of healing may become possible which would avoid the intervention both of drugs and surgery with their related dangers.

Psychic healing can actually involve what appear to be surgical operations. The practice of psychic surgery is particularly prevalent in the Philippines and South America. In a typical psychic operation the healer will seem to open up the body without the use of any surgical instruments, and may then remove pieces of tissue which are claimed to be the cause of the patient's illness. The opening is then closed, sometimes as soon as the healer's hands are withdrawn from inside the patient's body. In other cases the incision made in the skin will heal up over the normal period of time. Films of these psychic operations are hardly convincing so far as the sort of operation is concerned where openings in the body close up almost immediately and leave no scar behind. Some records give clear evidence of fraudulent practice. One film sequence of the removal of an eye to clean its socket had a clear picture of an eye with its optic stalk (the nerve fibres carrying information from eye to brain) cut and sticking straight out from the back of the eye. A freshly removed optic stalk, however, is not rigid and only after keeping it in a suitable preservative will it become so. Close scrutiny of the hands of the healer in other sequences suggests that they do not in fact penetrate the body. Whilst evidence of cheating does not mean that all such operations are fraudulent, they must obviously be regarded with suspicion.

There is another type of psychic operation and this has its practitioners throughout the world. Sometimes incisions are made in the skin without instruments, but the openings heal over in the expected length of time. At other times various illnesses are treated by the 'laying on of hands'. This kind is in a venerable tradition, Jesus Christ being the most famous of its exponents. If there is some truth in the many accounts of successful laying on of hands, how is it done? What is the healing field which some people possess and not others, and how does it operate in the sick patient?

We are wondering whether two of our children who are metal-benders may not also be psychic healers. The mother of one of them had severe pains in her hip over a long period of time and was taking a painkiller prescribed by her doctor, who said there was no other remedy. Her ten-year-old son laid his hand on her hip for a few seconds, during which time the mother felt a particular warmth in the deep tissue there. Within half an hour the pain had gone, and two months later it had not returned. And there is a similar report concerning a damaged knee.

Those who practise the laying on of hands remark that they sometimes feel a drain on their energy when healing in this manner. One healer even went so far as to say that some of his patients felt like 'sponges' absorbing his energy, the drain seeming to occur down his arms. What could be more natural than to suggest that it is electrical in nature, the healer stimulating tissue in the patient's body by the radiation he is emitting. This conjecture can be put to a clear test by measuring the radiation coming from healers during the healing process. It will be done in the near future and let us hope it will be successful. If so, it might lead to the building of a healing machine, which could perhaps intensify the effects produced by the human 'psychic' healer.

Let us finish with a word on miracles. The *Oxford English Dictionary* defines a miracle as a 'marvellous event due to some supernatural agency'. Invoking the supernatural is equivalent to saying that the event in question cannot be explained in terms of natural processes, that is, in terms of the coherent set of events which have clear causal relationships to one another. Thus the breaking of a window when a stone is thrown at it is not miraculous, though if the window broke when no stone or other object was hurled at it it might be regarded so. Throughout recorded history supernatural agencies have been invoked to explain the course of events. The many phenomena over which man has no power – storms, lightning and thunder, fire, fertility, growth, the movement of the planets in the heavens and the sun and moon – each had to have its particular god to control it. As the millennia passed these processes became explicable in terms of fewer basic ideas, helped especially by the enormous conceptual advance made by Democritus with his atomic hypothesis. All matter was constituted of indivisible atoms, their various types and motions producing the complexity of the world we see around us. This all-embracing hypothesis has served, especially in the last three hundred years, to explain an ever-increasing range of phenomena. Now atoms can be photographed, and their internal structure is understood.

Nevertheless so-called miracles have happened and still happen. But a miracle can only stay miraculous if it remains uninvestigatable. If its secrets become exposed under the penetrating gaze of science, its supernatural character will go. Very few miracles have been accessible to investigation and even fewer stood up to scientific analysis. Perhaps it will come about that only the miracles of the past will preserve their mystery.

Every so-called miracle is a challenge to science. The Geller effect is a case in point. Will it ever turn out that the miracles of Jesus Christ also dissolve in scientific explanation? It may well be hubris to hope that science can explain 'divine intervention', but to me it seems like cowardice not to take up the challenge.

This book has presented the case that for one modern 'miracle', the Geller effect, there *is* a rational, scientific explanation. This explanation is also claimed to allow us to understand other apparently miraculous phenomena – ghosts, poltergeists, mediumship and psychic healing. What, then, of other miracles? Can they too be explained by these newly discovered powers of the human body and mind, and the properties of matter broadly described in this book? Only time and the willingness to fly in the face of established belief will yield the answers. Clearly, there are more things in heaven and earth than are dreamt of in our philosophy.

Afterword, autumn 1975

The controversy about ESP still continues as I write these words. Those who had entrenched positions against the phenomena being out of the ordinary when this book first appeared still hold their same positions, with moderation in certain cases, exaggeration in others. In particular, metal-bending has produced a great deal of discussion, though sadly enough with more heat than light. Attempts have been made in various parts of the world to show that spoons are only bent by fraud, with demonstrations also claimed of *bona fide* paranormal spoon-bending by others. With such contradictions occurring it is clearly necessary to look at the evidence very carefully. Let us start with that.

Since he sparked off so much of the controversy, it is appropriate to commence with Uri Geller himself. Concern has been expressed, especially by conjurers, that Geller achieved his results by fraud. Various professional magicians have claimed that they can reproduce the spoon-bending and watch-starting and -stopping phenomena. One, James Randi, even convinced the editorial staff of the British parapsychological journal *Psychic Researcher* of his paranormal powers, as well as some of my colleagues at King's College in London (I was not present at the time). He at no time claimed he used paranormal methods to deceive, and has recently published a book describing how Geller managed his tricks.

Similar explanations of Geller's powers have also been published by the British popular science weekly, *New Scientist*, giving details as to how the frauds must have been achieved. Since Randi's book is not yet available I can only comment on the *New*

Scientist discussion. It concentrated on the telepathy tests carried out at the Stanford Research Institute and on the metal-bending tests at Birkbeck College, London University. In the former, various miniature radio transmitters and receivers were claimed to have been used to send the necessary information to Geller from an informant. There is the technical difficulty of the size of such devices, but assuming that they could have been so small as to be inserted, as it was claimed, in Geller's teeth (he has no cavities, nor ever visited a dentist, he recently told me) there is the major problem that the informant in certain cases would have had to be one of the scientific team involved in the test. Thus the charge of fraud could only stick if it included the scientists, and so has to be discounted unless all such evidence in this area is rejected. Even so, care must be taken here; no guarantee of impunity for cheating scientists can rest in the repeatability of their tests by other independent groups. The metal-bending tests at Birkbeck were criticized as being poorly controlled with too much confusion. But later tests appear to have avoided that difficulty, especially in the breaking of a crystal without direct contact by Geller, when Professor John Hasted claimed Geller could not have touched it at all.

Repeatability of metal-bending has occurred with Geller, at least at a certain level. Two tests have been reported on Geller which seem to have satisfied the people concerned, both with conjurer's training. One of these, W. E. Cox, reported tests of both key-bending and watch-starting. Two keys were bent by Geller, one of which was held down on a flat surface by Cox. In the other case, 'Geller had been handed this key, like the first, but returned it to me in a perfectly straight condition. My fore-finger pressed against the toothed end. Geller stroked only an inch of the handle, and this time a bend slowly appeared near that end, a full inch away from my finger. It conspicuously continued until it reached 36°. Again, there was no noticeable pressure upwards against my finger, and the time required was less than a minute.' Geller also started a watch which had been specially prepared. Cox's conclusion was that fraud could be ruled out, nor had the fellow magicians he consulted on this disagreed with him.

The other positive test was conducted by the magician Artur Zorka and one of his colleagues; in it Geller caused the nylon reinforced handle of a forged steel fork to explode when he touched it, and also accurately guessed several drawings made by the magicians or even only thought by them. They concluded '. . . there is no way, based on our present collective knowledge, that any method of trickery could have been used to produce these effects under the conditions to which Uri Geller was subjected.'

Naturally enough, the subject himself has something to add on this in his book *My Own Story*. It makes interesting reading, especially since the straightforward attitude presented, consistent with that Geller has shown in numerous media interviews, adds weight to the suggestion that he uses no deception.

The case for spoon-bending has been lent weight by the discovery of others, especially children, who have come forward claiming powers similar to Geller's. Here again deception occurs, though it can be spotted more readily than in the case of a practised magician. In my own investigations I have seen children who used methods which were somewhat suspicious, and Geller himself remarked on this to me. I have not continued working with cases of that sort as soon as the possibility of fraud has occurred. Recently two accounts of such deception have been reported, one by the *New Scientist* and the other by scientists from Bath University.

The *New Scientist* account was on the David Berglas/*Daily Express* spoon-bending competition (with a £5,000 prize given by Berglas). Several of the children were seen bending the spoons mechanically. Thus 'When attention was diverted to a child at one end of the room, a boy at the other end glanced around the room; confident that no one was watching he used two hands and all his strength to bend a fork.' The Bath investigation used one-way mirrors to observe a group of six children. Of these, five were caught clearly cheating; the sixth was unable to achieve anything.

The difficulty with the Bath tests was that no clear attempt was made to exclude fraud; thus the use of deception seemed to have been positively encouraged. For example, in the Bath tests 'the observers in the room were instructed to deliberately relax their vigilance at intervals after the first twenty minutes.' This also seems to have been the case with the Berglas/*Express* tests, for, as it was said, 'It was not possible to watch all the children constantly.' Even so, it was felt 'Some bent cutlery in a way not readily explicable by the observers.' This feature of childhood deception is clearly one to reckon with most carefully. As I said earlier, I had already experienced some of this, yet the evidence presented earlier in the book was taken in a way to guard against any such deception by child or adult. Tests were done either with the pressure being measured or contact being excluded completely by metal being placed in sealed containers.

Other scientists have found similar positive results. Thus Dr Sasaki and a group of fifteen researchers from Tokyo's Denki Tsuskin University found that some children could cause a metal wire inside a sealed tube to be bent downwards and also to cause other pieces of metal to bend. With the conventional cutlery even more drastic results occurred. Thus a twelve-year-old boy 'held three spoons, one at a time. The first two bent double almost immediately. The third spoon broke in two.' A scientist in Munich tested a thirteen-year-old girl, who caused a strip of metal he had placed flat on a desk to become bent upwards after she had gently stroked the top of it.

The situation with regard to spoon-bending is thus as controversial as ever, but the claims described above that it has been observed under fool-proof conditions would seem to outweigh those that it can also be achieved by fraud if close observation does not occur. The number of subjects who are able to achieve such results under the strict conditions necessary is not high; in my case I can guarantee no more than six. This is in agreement with the University of Bath results suggesting that no more than one in six of the cases claiming these powers would appear to possess them.

If we accept spoon-bending as occurring without fraud, is it a truly paranormal phenomenon or is there a natural explanation? In Chapter 8 I presented the only feasible explanation I could find which was still consistent with modern science, that it was caused by some low-frequency electromagnetic effect. Interestingly enough, metal can be distorted without contact by strong electromagnetic fields, and such an effect is utilized in what is called the 'electromagnetic hammer' which is used in the space programme. This achieves its effects by setting up stresses in metals by very strong magnetic material, stresses which exceed the yield strength of the material so that it flows in a plastic fashion yet is cold. We need to observe such strong fields around metal-benders if such is the basis of metal-benders' powers. Such fields are difficult to measure around them, though I have observed unexpectedly high ones during tests with a subject who was causing movement of objects. However, it may be that there are sudden electric field reversals due to charge transfer and it is these that cause the phenomenon. I have obtained results indicating that this could well be occurring, though these are still preliminary.

In any case, the problem is still to construct a machine which will achieve the same effects of metal-bending. That is still being pursued; hopefully with success within the next year or so.

Bibliography

Adey, W. R. *Evidence for Co-operation Mechanisms in the Susceptibility of Cerebral Tissue to Environmental and Intrinsic Electric Fields* (University of California, Los Angeles report)

Armstrong, D. M. *A Materialist Theory of the Mind* (Routledge and Kegan Paul, 1968)

Bawin, S. M., Gavales-Medici, R. J. and Adey, W. R. 'Effects of Modulated Very High Frequency Fields on Specific Brain Rhythms in Cats' (*Brain Research*, 58, pp 365–84, 1973)

Becker, R. O. 'Electromagnetic Forces and Life Processes' (*Technology Review*, 75, pp 32–8, 1972)

Bitter, Francis and Medicus, Heinrich A. *Fields and Particles* (Elsevier, 1973)

Borewits, Philip *Real Magic* (Sphere Books, 1972)

Carrington, Hereward *Eusapia Palladino and Her Phenomena* (T. Werner Laurie)

Cleary, S. F. 'Uncertainties in the Evaluation of the Biological Effects of Microwave and Radiofrequency Radiation' (*Health Physics*, 25, pp 387–404, 1973)

Collins, H. and Pamplin, B. 'Spoon-bending: an experimental approach' (*Nature*, 257, p 8, 1975)

Cottrell, A. H. *Dislocations and Plastic Flow in Crystals* (Clarendon Press, 1956)

Cox, W. E. 'A Scrutiny of Uri Geller' (American Parapsychological Convention 1974, 8/22 Brief [unpublished report], Institute of Parapsychology, Durham, North Carolina, USA)

Dooley, Anne *Every Wall a Door* (E. P. Dutton, 1974)

'The Electromagnetic Hammer' (Technology Utilization Report, NASA SP-5034, NASA)

Flew, Anthony *A New Approach to Psychical Research* (Watts, 1953)

Gauld, Alan *The Founders of Psychical Research* (Routledge and Kegan Paul, 1968)
Geller, Uri *My Own Story* (Robson Books, 1975)
Gowalas, R. J., Walter, D. O., Hamer, S. and Adey, W. R. 'Effect of Low-Level, Low-Frequency Electric Fields on EEG Behaviour in Macaca Nemestrina' (*Brain Research*, 18, pp 491–501, 1970)
Green, Andrew *Our Haunted Kingdom* (Fontana, 1974)
Hanlon, J. 'Uri Geller and Science' (*New Scientist*, 64, pp 170–89, 1974)
Hanlon, J. 'But What About the Children?' (*New Scientist*, p 567, 6 May 1975)
Harrop, P. J. *Dielectrics* (Butterworth, 1972)
Heywood, Rosalind *The Sixth Sense* (Chatto and Windus, 1959)
Hirth, J. P. and Lothe, J. *The Theory of Dislocations* (McGraw-Hill, 1968)
Hopkins, C. D. 'Electric Communication in Fish' (*American Scientist*, 62, pp 426–37, 1974)
Hughes, Pennethorne *Witchcraft* (Penguin Books, 1965)
'Japan's Incredible Psychic Children' (*National Enquirer*, 1975)
Johnson, C. C. 'Nonionizing Electromagnetic Wave Effects in Biological Materials and Systems' (*Proceedings of the IEE*, 60, pp 692–718, 1972)
Knudsen, E. I. 'Behavioural Thresholds to Electric Signals in High Frequency Electric Fish' (*Journal of Comparative Physiology*, 91, pp 333–53, 1974)
Koestler, A. *The Roots of Coincidence* (Hutchinson, 1972 and Picador, 1974)
Marha, K., Musil, J. and Tuka, H. *Electromagnetic Fields and the Life Environment* (San Francisco Press, 1970)
Marwick, Man (editor) *Witchcraft and Sorcery* (Penguin Modern Sociology Readings, 1970)
Mitchell, E. *Psychic Exploration* (Putnams, 1974)
Office of Telecommunications Policy Report on Program for Control of Electromagnetic Pollution of the Environment: *The Assessment of Biological Hazards of Non-Ionizing Radiation* (March 1973 and May 1974, Executive Office of the President, Washington, DC, USA)
Pearsall, Ronald *The Table Rappers* (Michael Joseph, 1972)
Presman, A. S. *Electromagnetic Fields and Life* (Plenum Press, 1970)
Puharich, A. *Uri* (W. H. Allen, 1974)

Puthoff, H. and Targ, R. 'Information Transmission under Conditions of Sensory Shielding' (*Nature*, 251, pp 602–7, October 1974)
Randi, J. *The Magic of Uri Geller* (Ballantine Books, NY, 1975)
Rhine, J. B. *The Research of the Mind* (Faber and Faber, 1954)
Rhine, J. B. *New World of the Mind* (Faber and Faber, 1954)
Rhine, J. B. and Pratt, J. G. *Parapsychology* (Blackwell Scientific)
Rhine, Louisa E. *Mind Over Matter* (Collier Macmillan, 1972)
Smythes, J. R. *Science and ESP* (Routledge and Kegan Paul, 1967)
Solymar, L. and Walsh, D. *Lectures on the Electrical Properties of Materials* (Clarendon Press, 1970)
Special Issue on Biological Effects of Microwaves, *IEE Transactions on Microwave Theory and Techniques*, February 1971
Summers, Montague *The History of Witchcraft and Demonology* (Routledge and Kegan Paul, 1973)
Taylor, J. G. *New Worlds in Physics* (Faber and Faber, 1974)
Taylor, J. G. *Black Holes* (Souvenir Press, 1973 and Fontana, 1974)
Taylor, J. G. *The Frontiers of Knowledge* (London Editions, 1975)
Taylor, J. G. *The Shape of Minds to Come* (Michael Joseph and Weybright & Talley, 1971; Panther, 1974)
Thonless, Robert H. *Experimental Psychical Research* (Penguin Books, 1963)
Tipper, C. F. *The Brittle Fracture Story* (Cambridge University Press, 1962)
Toynbee, Arnold (editor) *Man's Concern with Death* (Hodder and Stoughton, 1968)
Twigg, Ena *Ena Twigg, Medium* (W. H. Allen, 1973)
Tyler, P. E. 'Overview of the Biological Effects of Electromagnetic Radiation' (*IEE Transactions on Aerospace and Electronic Systems*, AES-9, pp 225–8, 1973)
Tyrell, G. N. M. *Apparitions* (Duckworth, 1943)
Vasiliev, L. L. *Experiments in Mental Suggestion* (Institute for the Study of Mental Images, Church Crookham, England, 1963)
Watson, Lyall *The Romeo Error* (Hodder & Stoughton, 1974)
Watson, Lyall *Supernature* (Hodder & Stoughton, 1973 and Coronet, 1974)
Weller, P. F. (editor) *Solid State Physics and Chemistry: an Introduction* (Marcel Dekker, 1973)
Wilkerson, T. E. *Minds, Brains and People* (Clarendon Press, 1974)

Zorka, A. 'Investigations of Uri Geller' (Report to Executive Committee, Society of American Magicians, Assembly 30, Atlanta, Georgia, USA unpublished)

Journals containing many papers on psychical research are:

Journal for the Society of Psychical Research (British Society for Psychical Research, 1 Adam and Eve Mews, London W8)
The Journal of Parapsychology (The Parapsychology Press, Durham, North Carolina, USA)
The Journal of Paraphysics (The Institute of Paraphysics, Downton, Wiltshire, England)
The Parasciences Journal (The Institute of Parasciences, Sprytown, Upton, Devon, England)

Index

Beyond Telepathy £1·25
Andrija Puharich

'Stories of normal men and women who have explored the far reaches of their minds . . . examples of telepathy, clairvoyance, and most remarkable of all – the personality freeing itself of the body and travelling where it will across the reaches of time and space' (from the Foreword by Andrija Puharich)

This is the classic work on the science of the paranormal. Andrija Puharich, a neurologist and the mentor and biographer of Uri Geller, gives a complete and convincing explanation of such phenomena as ESP and astral projection. And he presents a powerfully reasoned scientific justification for the expanded states of consciousness experienced in yoga, shamanism and possession.

The Roots of Coincidence 60p
Arthur Koestler

As research into parapsychology becomes more respectable scientifically, so the doctrines of modern physics become more and more 'supernatural'; here Arthur Koestler discusses several syntheses of physics and metaphysics. He finishes with a plea for open-mindedness in further research and a sternly practicable indictment of both rigid materialism and superstitious credulity. Argued in a clear and straightforward manner, this is a fine rambunctious essay in the punchy Koestler tradition.